Some "Prime" Comparisons

Some "Prime" Comparisons

STEPHEN I. BROWN
State University of New York at Buffalo

The National Council of Teachers of Mathematics

Third Printing 1991

Library of Congress Cataloging in Publication Data:
Brown, Stephen I.
 Some "prime" comparisons.
 Bibliography: p.
 1. Numbers, Prime. I. Title.
 QA246.B76   513'.2   78-12339
 ISBN 0-87353-131-0

Copyright © 1978 by
THE NATIONAL COUNCIL OF TEACHERS OF MATHEMATICS, INC.
1906 Association Drive, Reston, Virginia 22091
All rights reserved
Printed in the United States of America

This book is dedicated to

- my wife, Eileen Chaika,
  for our wedding anniversary
- our teenagers, Jordan and Sharon,
  for their birthdays

Though none of these anniversaries represented a prime (in N) during the year this book was published, they all were prime one year before and, I hope, will be so many times again.

Contents

PREFACE   ix

1. INTRODUCTION   1

1.1 An Overview 1
    A Silly Question? 1
    Intellectual Significance: How to Play with an Idea, 1
    The Focus and Use of this Book, 4

1.2 The Two Domains 6
    Problem Set, 7

1.3 Primes in N 9
    Problem Set, 9

1.4 Primes in E: A First Surprise 11
    Problem Set, 13

2. THE NUMBER AND THE GENERATION OF PRIMES   14

2.1 In N 14
    Euclid's Proof of the Infinitude of Primes, 15
    On Generating Primes, 16
    On Arithmetic Progressions and Infinitude of Primes, 20
    Summary, 23
    Problem Set, 23

2.2 In E 29
    A Closer Look at Generating Primes in E, 30
    On Revising the Generating Questions, 32
    Problem Set, 37

3. THE DETERMINATION AND DISTRIBUTION OF PRIMES   40

3.1 In N: Determination 40
   Is n a Prime? 41
      Summary of General Scheme, 45
      On the Meaning of "Workable," 52
      "Nontedious Approaches" in Special Cases, 48
      One-Shot Trial Procedures for All Numbers, 48
   Problem Set, 50

3.2 In N: Distribution 57
   The Primes between 1 and n, 57
      Making Things a Little More Elegant, 59
      Shifting the Question, 60
      On Integrating Eratosthenes and Hadamard, 62
      On Clustering and Doodling, 63
      A Helpful Digression, 67
      Back to the Spiral, 68
      In Defense of Doodling, 69
   Problem Set, 70

3.3 In E: Determination and Distribution 76
   Determination, 76
   Distribution, 78
      Eratosthenes Revisited, 78
      Hadamard Revisited, 79
      Ulam Revisited, 81
   Problem Set, 82

4. UNIQUE FACTORIZATION AND SURROUNDINGS   86

   A Recipe to Reduce (N and E), 86
   Behind It All (N and E), 88
      Unique Factorization, 88
      The Fall of the Greatest Common Divisor, 90
      Well, Order! 90
   How Nonunique Can You Be? (In E), 93
   Problem Set, 93

5. ODDS 'N EVENS   96

   Problem Set, 99

6. EPILOGUE   102

BIBLIOGRAPHY   106

Preface

The roots of this book go back a long way. In the summers of 1963 and 1964 I was responsible for the initial phases in the education of prospective teachers in the master's degree program at the Harvard Graduate School of Education. James Henkelman and I taught early versions of this material to junior high school youngsters and in that context attempted to introduce ten Harvard "interns" to the intricacies of teaching. The mathematical content of that first experience is discussed in my article in the *Mathematics Teacher* entitled "Of 'Prime' Concern: What Domain?" (1965). One of my interns, Edward Stone, also described some of the classroom interaction from the perspective of a beginning teacher (1965).

I subsequently extended these concepts and taught later versions of this material directly (without youngsters) to preservice teachers at Harvard and Syracuse universities in a course on methods of teaching mathematics. In addition, I taught it in the in-service program at State University of New York at Buffalo as a component in a course in number theory for teachers. Many of these students tested the material with their own secondary school youngsters and were kind enough to report back to me on its strengths and weaknesses.

Their criticism has enabled me to gear the tone of exposition as well as the sophistication of the exercises to intellectually alive high school students. As will be apparent in section 1.1, however, this book could also serve as a component in an introduction to number theory at the college level, as well as a "meta-text" for courses in teacher education. That is, if teachers read it and attempt to take one step back from the text, they will find embedded a number of essentially pedagogical issues dealing with how one learns mathematics. Some of this is mentioned further in the introduction after discussions on such dimensions as the significance of an idea and "pseudohistory." Andrew Berner, John Corcoran, Norton Levy, Edwin Moïse, Gerald Rising, and Marion Walter have criticized some of the original ideas or have been influential in helping to generate and criticize some of the pedagogical insights behind these dimensions.

To them, to James Henkelman and my interns, who assisted me in embarking on this venture, and to my many students at all levels who shared the excitement of learning it or who tested it with their own students, I am extremely grateful. In addition, Betty Krist, George Goodwin, and Elias Saba, my graduate assistants; Debbie Tricoli and Mary Sullivan, my typists; and Charles Clements and Nicholas Ronalds, of the NCTM editorial staff, were of great help in preparing the manuscript for publication.

Introduction

1.1: AN OVERVIEW

A Silly Question?

CAN YOU reduce the fraction 12/36 to lowest terms? Without too much calculation, we realize that 12 divides both numerator and denominator, and therefore 1/3 is our answer. Can you reduce the fraction 12/36 to lowest terms so as to yield an answer different from 1/3? What a peculiar question! Since 1 and 3 are relatively prime, we have gone as far as we can go. That's true, but could we perhaps have found other factors common to 12 and 36 such that the reduced-to-lowest-terms fraction might equal 1/3 but have a different numerator and denominator? Our experience with reducing fractions to lowest terms tells us that such things are not possible. But what do we understand from this experience? Might things be otherwise? Let's digress briefly from this question in order to place it in a broader perspective.

Intellectual Significance: How Do We Play with an Idea?

It is one thing to believe that something is so (by deducing it from other assumptions, by using inductive inference, or perhaps by some other means). It is another thing to see the intellectual significance of our understanding, knowledge, or belief. Using topics from elementary number theory dealing with prime numbers, we have selected this issue—that is, of how to learn to appreciate the intellectual significance of our mathematical understanding—as an important focus of our exploration.

What does it mean to say that we are more concerned with conveying the intellectual significance of an idea than with offering mathematically sophisticated proofs? Perhaps an anecdote will point up the distinction.

A colleague told me of his experience of sitting in on an oral examination of a student who was reporting on his doctoral dissertation findings—the last stage of a supposedly serious intellectual investigation on a branch of algebra, Banach algebra. One of the examiners asked the student to cite an example of the general theory he had investigated. Only after much prodding from the examiners was he able to come up with even a single, trivial example of the theory about which he had spent at least a year thinking seriously.

What was missing? It was not that the doctoral student had done an inferior job of proving anything, but rather that he had almost no idea of why he was proving things in the first place. In short, though he was answering a question that someone may have felt was significant, he had little appreciation himself for why it was significant.

To understand the significance of an idea, we need to engage in all kinds of activities in addition to (or even instead of, on occasion) demonstrating proofs. People interpret psychologists like Piaget to be saying that play with concrete material is essential. But this is a partial and misleading truth, for it is not the concreteness of the material but rather the playing with ideas that is essential if we are to understand the significance of our investigation.

How do we play with ideas? We do so by learning to make endless variations of the phenomenon we are investigating. In this book the beginning of a variation on the set of natural numbers and primes is suggested by exploring what happens when we tamper with that set in what appears to be minor ways.

But what is involved with tampering? It is on the one hand a matter of varying the context in which we perceive a thing and on the other a matter of holding a context fixed and varying the domain we investigate. Of course, varying the domain of investigation will accomplish very little in itself unless we engage in another important activity: asking questions and analyzing the answers to these questions in their new contexts. We gain even greater insight by comparing answers from different contexts.

So, for example, how do we learn, not the proof, but the *significance* of the fundamental theorem of arithmetic—that any number can be factored into a product of primes in exactly one way? We do so, not by proving the theorem alone, but by seeing what happens to the fundamental theorem of arithmetic when we modify the context. We might ask, for example (if we modify our original system slightly so that we no longer focus on natural numbers), What happens to the fundamental theorem of arithmetic? We might also investigate, in this modified system, what happens to the reducing of fractions to lowest terms. One interesting discovery will be that there are systems that have many different reductions of fractions to lowest terms. We shall find a system in which $12/36$ can be reduced to lowest terms in two different ways. We shall then have totally new lenses through which to see the uniqueness in the reduction of fractions to lowest terms in the set of natural numbers. It will then be less of a humdrum question to ask why 12/36 can be reduced to lowest terms in only one way.

Given a system in which some of our cherished "self-evident" truths fail, we have an entirely new motivation to examine proofs that assert these truths in our familiar system. We now have considerable motivation to search for points at which attempted proofs fail as we switch from one system to another, and we begin to see for the first time the power of previously unrecognized assumptions.

In our system of mathematics education very little time is spent turning ideas inside out to discover how the answers to questions vary even with minor modifications of the idea under investigation. We shall see in this book, for example, how it is that questions unanswerable in one context become answerable in others. We shall also see how questions foolish in one context become profound in another.

Another heuristic for understanding the significance of the ideas we investigate is to make use of what is here called "pseudohistory." (It should be noted that the author is not a historian and cannot claim to be able to trace accurately the evolution of concepts presented here, but neither he nor the student concerned with understanding the significance of an idea need necessarily be able to do so.)

Moreover, it is also helpful to have an ability to ask, not "How *did* this idea come about?" but rather "How *might* this idea have come about?" That and questions of the following type help to give us insights into the significance of an idea:

1. What do you think were some of the major difficulties people had when first attempting to state a particular proposition?
2. What do you think were some of the major difficulties in proving or disproving the proposition?
3. What issues might have motivated people even to investigate the idea in the first place?
4. What do you think a rudimentary version of this idea might have been a thousand years ago?

Let us turn to a brief example of what seems to be lacking in terms of the significance of ideas. Transformation geometry is becoming an important part of the elementary and the secondary mathematics curricula. After an introduction to the meaning of symmetry and motion, most books encourage us to investigate four different types of motion—reflection, translation, rotation, and glide reflection. We are then given many exercises that enable us to become proficient (1) in distinguishing the four types of motion, either intuitively or rigorously, (2) in producing them, and (3) in relating them to each other.

So far, so good. What is missing, however, is some questioning and some exploration that might bring home the *significance* of these four motions. Why have these four been singled out? Certainly in part because one can show "nonmessy" relationships among them; and in part because one can derive from them important conclusions that do not have these motions in the foreground. That is fine, but how do we find out how unmessy, and therefore elegant, these four motions are? It is not just by proving the theorems that interrelate them that we come to such understandings; it is also by examining the consequences of selecting other motions. There are certainly an infinite number of nonconventional ways in which we can move from a to b (some even involving motion in three-space). What, then, are the consequences of selecting some nonconventional motions? How might the conventional ones have come about? Unless we begin to ask questions of this kind, our investigation will degenerate into a "rhetoric of conclusions" (to use a phrase popularized by Joseph Schwab in his criticism of the modern curriculum) rather than launch us on a mind-expanding journey.

The Focus and Use of This Book

As our vehicle for encouraging a mind-expanding journey, we introduce two different domains, $N = \{1, 2, 3, 4, \ldots\}$ and $E = \{1, 2, 4, 6, 8, 10, \ldots\}$. Comparing these two domains (and occasionally others), we begin to see how answers to questions are extremely sensitive to modifications in the phenomenon investigated. And just as important, we begin to see the need for a clarification of questions when, for example, we obtain two different answers to the question of what is a prime in E, depending on how we interpret the concept of prime in this new setting. Furthermore, one interpretation of prime in E not only enables us to answer questions about primes in E but also allows us to see in a clearer light what is involved in the concept of prime in the familiar N.

Though we shall assume that the reader understands the concept of the variable and can follow and create elementary proofs involving it (including proofs by contradiction), the presentation is "gentle" in terms of the pains taken to introduce terminology, concepts, and symbolism (such as the principles of arithmetic, the notion of a constructive proof, factorial notation, etc.) that may have been forgotten. In addition, great care has been taken to integrate exercises with textual explanation. (Exercises are denoted by x.y.z., where x is the chapter in which the exercise appears, y the section, and z the number of the exercise in the section.)

An aspect of the book's format adds to its "gentleness." Because of the constant comparison of N with E, the reader does have the option of skimming parts of the discussion of N that he or she finds difficult and of turning to the analogous issues in E. After so gaining an extended picture of the terrain, the reader can return to the discussion in N with greater clarity. What we are making is essentially a forest-and-tree distinction. Too much mathematical exposition (and learning in school in general) produces a linear rather than a gestalt model of learning uncongenial to the large number of people whose approach to the world is more holistic.

In some ways, however, the presentation here is more challenging than that of most introductory texts in number theory. Because of its emphasis on pseudohistory, we do presuppose an alert mind and a willingness on the part of the student to pursue some issues with the curiosity and honesty that must have characterized many of those who first embarked on such investigations. In short, we attempt—without requiring a lot of knowledge or mathematical sophistication—to re-create the excitement of original investigation that is so often lacking in "polished" works of mathematics, works that give the illusion of having been born full blown, like Athena from Zeus's head (with no labor pains to boot!). With regard to the inability of mathematicians to come up with a simple generating formula to produce all the primes, for example, we encourage the reader to whittle away at the question until some weaker but satisfactory question and analysis are found.

Because of our emphasis throughout on the significance of an idea as well as on the comparison of similar questions in different contexts, this exploration has a philosophical side that requires thoughtful input from the reader. The reader is asked not only to answer questions but to analyze what it means for a question to be answerable. We ask questions that appear on the surface to be foolish: How do you know that there's no divisor of 3599 in N that is bigger than 3599? For such questions we beg indulgence, for it is frequently the most foolish-sounding and trivial questions that generate the kind of wonder that is a quantum leap beyond that generated by explorations of the moon!

We mentioned in the Preface that this book might be useful as a "meta-text" for teachers. Let us sketch how that might be accomplished. In addition to attempting the exercises, teachers might find it profitable to reflect on the *nature* of the activities suggested in the text and problem sets. For example, there are many open-ended exercises ("Explore what happens if . . ."); there are exercises that suggest that students debate each other about apparent paradoxes; there are exercises that suggest cooperative effort among readers; and so forth.

Teachers could profit from exploring the assumptions behind these exercises as well as their appropriateness for people at different levels of sophistication. In addition, teachers might do the following:

1. Apply the fundamental pedagogical ideas (pseudohistory and notions of significance and modification of perspective) to other topics in the school curriculum.
2. Create additional exercises for the book that will enlarge its scope. There are, for example, a number of places where concrete materials

(such as Cuisenaire rods) are suggested. What are additional places where such material can be used? What fundamental assumptions that have been glossed over require delicate machinery to be taught? Can new domains be created for many of the exercises, and if so, how do they compare with those in the text?

1.2: THE TWO DOMAINS

The two settings that will provide a backdrop for the major theme of the book are (1) the set N of natural numbers, $\{1, 2, 3, 4, 5, 6, 7, 8, \ldots\}$, and (2) the set E, which consists of 1 together with all the even numbers of N, that is, $\{1, 2, 4, 6, 8, 10, 12, 14, \ldots\}$. In addition, a set that will be used in the problem sets is $T = \{1, 3, 6, 9, 12, 15, 18, \ldots\}$. The major theme—that of prime numbers—we shall encounter soon. In this section we shall learn a little about the backdrops. You may be curious about why it is that we have included 1 as an element of E and T. What does 1 add to the set of even numbers and to multiples of 3 that motivates us to include it? A clue to the answer will be suggested in the following section.

Since we shall on occasion want to focus on these sets (E and T) with 1 deleted, let us for convenience create E' and T':

$E' = \{2, 4, 6, 8, 10, 12, 14, \ldots\}$, the set of even numbers of N

$T' = \{3, 6, 9, 12, 15, \ldots\}$, the set of multiples of 3

Let us, for a moment, look at one important property of E'. If we take any two elements of E' and add them, what is the result? A few specific examples should persuade you that we always end up with an element of E'. Is it always true that if we add two even numbers (elements of E'), we get an even number (element of E')? One way of further persuading yourself that this is so is by glancing at figures 1 and 2.

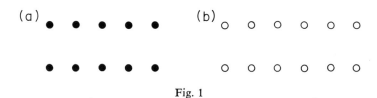

Fig. 1

To claim that a number is even (belongs in E') is to claim that it is divisible by 2 in the set N. How do we demonstrate that a number is divisible by 2 schematically? One obvious answer is that from it we can form two piles of exactly the same number of elements. In figure 1, pile (a) and pile (b) denote two numbers, both divisible by 2. To show that their sum is also divisible by 2, merely slip pile (b) next to pile (a) (see fig. 2), and

Fig. 2. A model to demonstrate that the sum of two evens is even

note that we can form two piles with an equal number of elements in them (try ex. 1.2.2). In figures 1 and 2 marbles are used to indicate the total number of elements. They could just as well be depicted with material such as Cuisenaire rods—sticks of different colors and lengths that can represent the numbers one through ten.

Since the sum of two even numbers in N is also even, we claim that the set E' is *closed* with respect to addition. In general, to say that a set is closed with respect to an operation means that the element obtained after performing an operation (like adding) on any two elements of the original set belongs to that same set (try ex. 1.2.1).

A more formal argument that the set E' is closed than the one presented in this section can be found in exercise 1.2.3. Since three basic ideas that will be used throughout this text are interrelated in that argument, you might wish to try that exercise now.

Though E' is closed with respect to addition, it is obvious that E is not. Why? Because both 1 and 2 are elements of E but $1 + 2$ is not an element of E. Can you find another example to show that closure breaks down? Obviously, from the point of view of closure under addition, E is less appealing than E'. We shall soon discuss why we are interested in sets such as E, however.

PROBLEM SET 1.2

1. Which of the following sets is closed under the indicated operation? If closure fails, give one example to show that it does.

   a) N under addition
   b) N under subtraction
   c) N under division
   d) N under multiplication
   e) The set consists of the following colors: red, yellow, and blue. Let the operation be, Mix any two colors.
   f) Consider the set of all roses. Let the operation be, Pollinate two healthy roses (i.e., roses capable of bearing flowers).
   g) The set consists of all the colors that can possibly be produced. Let the operation be, Mix any two colors.

2. In figures 1 and 2, we showed schematically why it is that if two numbers are divisible by 2, then so is their sum. Demonstrate schematically why the same would be true for two numbers divisible by 3.

3. Below is a more formal proof that the sum of two numbers is divisible by 2 if each of the numbers is so divisible. We shall make use of the following three basic ideas:

   i) N is closed under addition.
   ii) If a number is even, it can be expressed in the form $2 \cdot n$ for some n belonging to N.
   iii) For any a, b, c belonging to N, $a \cdot (b + c) = a \cdot b + a \cdot c$. This is called the distributive principle.

   Proof. If x and y are any two even numbers, then $x = 2 \cdot j$ and $y = 2 \cdot k$ for some j, k in N, by (ii) above. Furthermore, $x + y = 2 \cdot j + 2 \cdot k = 2 \cdot (j + k)$, by (iii), but because of (i), $(j + k)$ belongs to N. Let $j + k = f$. Since $x + y = 2 \cdot f$ for some f in N, $x + y$ must be even, by (ii).

   Using the principles above, show that T' is closed with respect to addition, that is, show that if any two numbers are divisible by 3, then so is their sum. Can you generalize?

4. Is the set of odd numbers closed under addition? Use marbles or Cuisenaire rods to show what happens whenever you add two odd numbers. Arrange the marbles so that you have a general proof, even though you are demonstrating it for only one case. Is it possible to do this?

5. What can you say about the product of two elements of N? E? E'? T? T'? Prompted by the definition of *even* in exercise 3(ii) above, prove your conclusion for E' and T'.

6. For (ii) through (vii) below, try to give a formal proof, as in exercise 3 above. For which of these can you demonstrate what is going on by using marbles or Cuisenaire rods? We claimed that any even number in N is of the form $2 \times n$ for n belonging to N. (In this exercise we introduce a notation that will be used throughout this book. If x is an element of a set A, we shall write $x \in A$, meaning x *belongs to* or *is an element of A*.)

   i) How could you express any *odd* number in N? (*Hint:* Given any even number, how would you find the nearest odd number that is less than it?)
   ii) Prove that the product of two odd numbers is odd.
   iii) Prove that E' is closed with respect to multiplication.
   iv) Prove that the product of an odd and an even is even.
   v) Prove that no odd number is divisible by an even number.

vi) Prove that if $n^2 \in E'$, then $n \in E'$. (*Hint:* Try an indirect proof.)

vii) Prove that if $n^2 \in N$ is odd, then so is n. (*Hint:* Try an indirect proof.)

1.3: PRIMES IN *N*

We now turn to the main theme of the book: prime numbers. A number is prime in N if it has exactly two different factors (the chart below should remind you what is meant by factor in N). Consider the first twelve elements of N together with their sets of factors:

N	Set of Factors
1	{1}
2	{1, 2}
3	{1, 3}
4	{1, 2, 4}
5	{1, 5}
6	{1, 2, 3, 6}
7	{1, 7}
8	{1, 2, 4, 8}
9	{1, 3, 9}
10	{1, 2, 5, 10}
11	{1, 11}
12	{1, 2, 3, 4, 6, 12}

The first five primes in N are 2, 3, 5, 7, and 11. Notice that our definition clearly excludes 1 as a prime. You probably will be able to do the exercises in problem set 1.3 without being given a formal definition of *factor* (also called *divisor*). Should you have any difficulty, however, you might wish to glance at section 1.4, where (motivated by the desire to find primes in E) we offer a formal definition.

PROBLEM SET 1.3

1. What are the next two primes in N after 11?

2. If p is a prime number in N, what is its set of factors?

3. Why doesn't 1 satisfy our definition of prime?

4. Some numbers have exactly three factors. Two of them are 4 and 9. What are the next two numbers of N that have exactly three different factors? Can you describe all numbers of N that have this property?

5. Which of the following are primes in N (and why)?

a) 24×15

b) 23×17

c) 3599
d) $2^2 + 2 + 41$
e) $3^2 + 3 + 41$
f) $4^2 + 4 + 41$
g) $41^2 + 41 + 41$ (*Hint*: Try to solve this without tedious calculation.)
h) 10^{1330} (*Note:* This is a large number. How many digits does it have?)

6. If two numbers x and y in N have a common factor m, how can you represent x and y in terms of m? (Recall how you represented even numbers. See ex. 1.2.3.)

7. Clearly, 2 is prime because it has exactly two factors, 1 and 2. Prove that there are no other even primes in N.

8. Use either standard graph paper or Cuisenaire rods for this exercise, the purpose of which is to suggest in a more intuitive way what a prime number is. Choose the smallest square (or the smallest rod) as a unit representing the number 1.

a) How many different-shaped rectangles can you make with this unit?

b) If you use two unit squares, how many different-shaped rectangles can you make? (You may place two squares so they abut but do not overlap.)

c) If you use six unit squares, how many different-shaped rectangles can you make?

d) Complete the following table:

Number of Unit Squares	Number of Different-Shaped Rectangles
1	
2	
3	1
4	
5	
6	
7	
8	2
9	
10	
11	
12	

e) For eight units, the two ways are as follows:

i)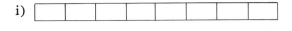

ii)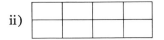

Why don't we consider the following to be a different way?

f) Compare your table with that at the beginning of section 1.3. What do you notice?

1.4: PRIMES IN *E*: A FIRST SURPRISE

Recall that $E = \{1, 2, 4, 6, 8, 10, 12, 14, \ldots\}$. What are the first few primes in *E*? The question is deceptively easy. Our first inclination is to claim that 2 is the *only* prime in *E*, since no even numbers in *N* beyond 2 are prime in *N* (see ex. 1.3). We might claim, for example, that 6 is not prime in *E*, for it is divisible not only by 1 and 6 but by 2 and 3 as well. This is not so, however, and to see where the problem lies, let us reexamine an analogous problem in *N*. We have claimed that in *N*, 5 is prime because it has exactly two different factors, 1 and 5.

One might (erroneously) wish to claim that there are factors of 5 other than 1 and 5. For example, isn't 5/2 a factor of 5? It looks as though we are claiming that if we can find two numbers *a* and *b* such that $a \cdot b = c$, then *a* and *b* are both factors of *c*. Would it therefore be legitimate to claim that 5/2 is a factor of 5, since $\frac{5}{2} \times 2 = 5$? By such reasoning 2 would also be a factor of 5. We could, in fact, get any number of factors of 5, since $5/3 \times 3$ and $5/17 \times 17$ and $2/3 \times 15/2$ all equal 5. Why, then, don't we claim that these numbers are factors of 5?

Looking back at the table of factors in section 1.3, we might immediately respond that "fractions are not allowed." This is true but misleading; let

11

us try to generalize our observations so that we can apply our insights to the concept of prime in many sets. Why is it that fractions are not allowed? The answer is that in order to claim that a is a *factor* (also called a *divisor*) of c in the set of natural numbers N, we require that—

1. $a \in N$;
2. there is another number b also belonging to N such that $a \cdot b = c$.

Now can we say why 5/2 is not a factor of 5 in N? Which of the two conditions above is not met? Since 5/2 does *not* belong to N (why?), obviously condition (1) is not met.

How about 7? Is 7 a factor of 5 in N? Certainly 7 belongs to N; so if we can find another number by which to multiply 7 to get 5, wouldn't 7 be a factor of 5? There is such a number—5/7—since $7 \times 5/7 = 5$. Then what's the problem? The answer is that condition (2) above is not satisfied, that is, 5/7 does not belong to N. Only condition (1) is satisfied for the number 7 in N.

Let us now return to the set E. What are the factors (or divisors) of 6? Certainly 1 is a factor of 6 because—

1. $1 \in E$;
2. there is some other number—6—belonging to E, such that $1 \times 6 = 6$. For the same reason, 6 is a factor of 6.

And what about 2? Is 2 a factor of 6? The first question we have to ask is, In what domain? In N 2 is a factor of 6 because (1) $2 \in N$ and (2) there is another number—3—belonging to N, such that $2 \times 3 = 6$. What happens in domain E, however? Does 2 divide 6 there? It is true that $2 \in E$. What about the second requirement for an element to be a factor of E? Is there *some other number (x) belonging to E* such that $2 \cdot x = 6$? Our immediate inclination would be to say, "Of course there is: $x = 3$." But the problem is that 3 does *not* belong to E, just as 2 was not a factor of 5 in N because 5/2 does not belong to N.

With the definition of factor above clearly in mind, let us list the factors of the first few elements in E:

Elements of E	Factors of Elements of E
1	{1}
2	{1, 2}
4	{1, 2, 4}
6	{1, 6}
8	{1, 2, 4, 8}
10	{1, 10}

The definition of prime alone prompts certain questions concerning the number and distribution of primes. Other questions—concerning divisibility —arise only after further probing. Though aspects of number theory are interrelated, we shall for convenience break down the analysis according to the following categories: Number and the Generation of Primes (chap. 2); Distribution and Determination (chap. 3); Unique Factorization and Surroundings (chap. 4); and Odds 'N Evens (chap. 5). Some overall structural issues are discussed in the Epilogue (chap. 6.)

PROBLEM SET 1.4

1. Find the next two primes in E beyond 10.

2. If p is a prime in E, then what are its factors?

3. Recall the set $T = \{1, 3, 6, 9, 12, 15, \ldots\}$
   a) Does 3 divide 6 in this set? Why?
   b) Does 2 divide 6 in this set? Why?
   c) Is 6 prime or composite? Why?

4. Let Q be the set of positive rational numbers (which includes the counting numbers as well: 1/9, 2/7, 13/59, 82/5, 3, 18, 4, and so on). Consider the number 5/23. Is 23 a factor of 5/23? Is 5 a factor of 5/23? Is 1/5 a factor of 5/23? Is 17 a factor of 5/23? List one element of Q that is *not* a factor of 5/23. Answer the same set of questions for *any* number belonging to Q other than 5/23.

5. List any two primes of Q.

6. The statement "a divides b" is another way of saying that a is a factor of b. A symbolic representation for both statements is "$a|b$." Now consider a domain K. How do you determine whether $a|b$ in K?

7. Look once more at exercise 1.3.8, which attempted to give you a feeling for prime in N. It turns out that a number n is prime in N if only one rectangle can be made with n unit blocks. If we want to choose this same intuitive definition for prime in E, how can we arrange our rectangles so that 6, for example, *cannot* be formed as follows:

Answer the same question with 10 and 14. That is, can you think of some scheme that would tell you when stacking is possible in E?

2

The Number and the Generation of Primes

2.1: IN *N*

A GLANCE at table 1 shows that the number of primes tends to become more sparse as we progress through the set of natural numbers. There are twenty-five prime numbers between 1 and 100, for example, but only fourteen between 901 and 1000. We might very well suspect that after a certain point, there are no more prime numbers in N. If this were so, many of the questions we shall explore in this chapter would lose their attraction for us (or at the very least we should have to modify them).

Over two thousand years ago Euclid came up with an answer to our question of whether primes no longer occur beyond a certain point. His proof that there are an infinite number of primes is very brief and one of the most elegant of mathematical proofs. Though brief, the proof is a little slippery, and you may have the feeling that you see it one moment and lose it the next. If you do not understand it after one or two readings, you might wish to push on through the rest of the chapter and return to the

TABLE 1

The Distribution of Primes in the First 1000 Elements of N

Range	Number of Primes
1–100	25
101–200	21
201–300	16
301–400	16
401–500	17
501–600	14
601–700	16
701–800	14
801–900	15
901–1000	14

proof later on. Exercise 2.1.1, which attempts to clarify a widely held misconception regarding the proof, may also be helpful. Only exercises 2.1.13–2.1.16, concerning the work of Dirichlet, and exercise 2.1.17, concerning an issue we discuss in the subsection following the next one, require an understanding of Euclid's proof to be appreciated. You might profit from reviewing Euclid's proof just before tackling those exercises.

Euclid's Proof of the Infinitude of Primes

Let us now look at Euclid's proof of the infinitude of primes. It is a proof by contradiction. Euclid assumes that there are a finite number of primes and shows that this assumption leads to difficulties. The first few primes are 2, 3, 5, 7, 11, 13. Let us denote the first one by p_1 (so $p_1 = 2$), the second by p_2 ($p_2 = 3$), the third by p_3 ($p_3 = ?$), and so on. If we assume that there are a finite number of primes, then we can arrange them in ascending order until we reach the last one. Let these primes be denoted by $p_1, p_2, p_3, p_4, p_5, \ldots, p_n$, where p_n is the last prime.

Euclid's proof now suggests that we form a new number and look at it closely. What motivates us to create this particular number is something you will appreciate as you go through (and review) the proof. The new number, K_n, is formed by multiplying together the finite list of primes and then adding 1:

$$K_n = (p_1 \cdot p_2 \cdot p_3 \cdot \ldots \cdot p_{n-1} \cdot p_n) + 1$$

As you will see when you understand the proof, it took an act of supreme genius to realize that the creation of this number would be helpful. This number is certainly bigger than any one of the enumerated prime numbers. Why? Let us look closely at K_n. There are two possibilities with regard to K_n and the concept of prime: either K_n is prime or K_n is not prime. Let us review each of these possibilities.

1. Suppose K_n is prime. If so, we have formed a new prime (K_n itself) larger than any of the previously enumerated primes $p_1, p_2, \ldots, p_n$, and we therefore cannot claim that $p_1, p_2, \ldots, p_n$ exhausts all the possible primes.

2. If K_n is not prime (and we call numbers that are neither 1 nor prime composite), then it must be a product of primes and hence divisible by *some* prime. (This should not be hard to see intuitively. You might wish to do exercise 2.1.4 if you would like to see a formal proof.) But K_n cannot be divisible by p_1, for

$$\frac{p_1 \cdot p_2 \cdot p_3 \cdot \ldots \cdot p_n + 1}{p_1} = p_2 \cdot p_3 \cdot \ldots \cdot p_n + \frac{1}{p_1},$$

and we thus have a remainder of 1. For the same reason, K_n cannot be divisible by p_2. The same claim can be made for $p_3, p_4, \ldots, p_n$.

Therefore there must exist *some* prime other than $p_1, p_2, \ldots, p_n$ that divides K_n, and so we are forced to conclude that $p_1, p_2, p_3, \ldots, p_n$ cannot exhaust all the primes.

A first hasty reading of this proof might suggest that we are claiming that K_n must always be prime (or perhaps that it must always be composite). Notice we make no claims about what K_n *must* be. We are merely showing that if $p_1, p_2, \ldots, p_n$ exhausts the primes in N, then when we consider the nature of K_n (and there are two alternatives), we reach a contradiction. You might now wish to try exercise 2.1.1.

On Generating Primes

Realizing that the primes are infinite in number, we might at first ask, Is there a formula that yields all the prime numbers (even allowing for the possibility that they may be generated "out of order")?

The (vain) search for an answer to this innocent question has consumed the energy of brilliant mathematicians for centuries. Even the attempt to answer the following less demanding question has been unsuccessful (though we shall presently reveal a peculiar sense in which it has recently been "solved"): Is there a formula that generates an infinite number of primes (and only primes) but not necessarily every prime?

Prior to the age of calculators and computers, conjecturing about the generation of prime numbers was a relatively safe business. The conjectured primes generated by proposed formulas were so large after the first few and the "brute force" calculation needed to determine the primality of large numbers was so inefficient that it was extremely difficult to disprove even some rather simple and incorrect conjectures. It often took a century or more to debunk such formulas.

Attempting to answer the second, less demanding question above, Fermat (1601–1665) conjectured that all numbers of the form $2^{(2^n)} + 1$ for n a natural number yielded only primes. If we denote a Fermat number by $F(n)$, then by plugging numbers 1 through 5 into $F(n)$, we get $F(1) = 5; F(2) = 17; F(3) = 257; F(4) = 65\,537; F(5) = 4\,294\,967\,297$.

It is obvious without much calculating that 5, 17, and 257 are primes. Perhaps Fermat has hit on a formula that will always work. What about 65 537 and 4 294 967 297? Are they primes? How would you go about finding out? Though we shall presently attempt to answer these questions, clearly the direct method of determining primality, dividing each number by numbers less than itself, is quite a laborious task. It took the genius of Euler (1707–1783), approximately a century after Fermat's conjecture, to show that $F(5)$ is composite (it is divisible by 641). His was a laborious route to follow in his time.

Mersenne (1583–1648) conjectured that all numbers of the form $2^p - 1$ for p itself a prime would generate primes. Thus $2^2 - 1 = 3; 2^3 - 1 = 7;$

$2^5 - 1 = 31$; $2^7 - 1 = 127$. All these numbers are prime. As with Fermat, Mersenne's prime number generator, $2^p - 1$ for p a prime, breaks down for the fifth value of p: $2^{11} - 1 = 2047 = 23 \cdot 89$. (See ex. 2.1.6.)

We indicated earlier that in a peculiar sense the problem of discovering a formula that would generate an infinitude of primes (allowing for omissions) was solved. In 1947, W. H. Mills (1947) proved that there is a real number A having the property that the *largest integer not exceeding* A^{3^n} (denoted $[A^{3^n}]$) is a prime for every natural number n (see ex. 2.1.7 for a clarification of the meaning of "[ ]"). The peculiar aspect of the solution is that Mills was not, in his proof, able to determine either the specific or even the approximate value of A. Very often in mathematics one has to resort to what is known as a nonconstructive proof, that is, a proof with no indication of how to locate an element having a desired property, even though we know that such an element must exist.

We have already come across a nonconstructive proof in this book. Recall that in Euclid's proof of the infinitude of primes, we showed (in case 2) that if K_n is composite, there must exist a prime that is different from $p_1, p_2, p_3, \ldots, p_n$. The proof alone involves no way of actually producing that prime (see ex. 2.1.17).

It is a source of great amazement that humans have learned to create a way of proving beyond a shadow of a doubt that something must exist without expressing in any precise (or for that matter imprecise, as with Mills) manner what the value of that something is.

An interesting aside here concerns what is known as the intuitionist school of mathematics. The ideas of this school were originated by Poincaré 1854–1912) and Kronecker (1823–1891) and formalized by the Dutch mathematician Brouwer (1881–1963) around the turn of this century. The school's philosophy holds that an entity whose existence is to be demonstrated must be constructed in a finite number of steps based on (the obvious) properties of the "God given" set of natural numbers. Many of the theorems related to primes in N require proof by contradiction, as in Euclid's proof of the infinitude of primes, which would not be acceptable to the intuitionist school. If we accept their position, the source of amazement we have been discussing is thus given the ax.

As we proceed in this book to furnish proofs in E (as we compare N and E), you may find it enlightening to ask how many of the proofs would in fact be acceptable to the intuitionist school.

Brilliant as Mills's contribution is (and we have suggested no hint of his strategy of proof), it is nevertheless true that for all practical purposes (since the formula does not enable us to specify what prime $[A^{3^n}]$ is for any n—though we know it must be *some* prime) no algebraic formula has been produced that will generate an infinite number of primes and *only primes* without necessarily generating *every* prime. That is, we have not received an affirmative answer to our second, less demanding question.

What are the consequences of once more relaxing the criteria of the desired formula? Suppose we allow the intrusion of many nonprimes in the sequence. We are thus led to consider the following question: Can we suggest a formula that will generate an infinite number of primes allowing—

a) the omission of many from the sequence;

b) the inclusion of many nonprimes as well?

As we have seen, both Mersenne and Fermat proposed formulas that generate both primes and composites, and it was by default rather than intention that composites appeared in the sequence at all. At this time, however, it is not known whether or not the formulas of Fermat and Mersenne meet even the relaxed criteria for the generation of an infinite number of prime numbers cited above. It may be that after some point their formulas generate only composites. In the absence of a simple and efficient procedure for determining primality (to be discussed in the following chapter), it is not surprising that we cannot crack such innocent-looking formulas to determine whether or not primes fail to appear after a certain point in the sequence. Thanks to modern digital computers, we now know that the Mersenne number $2^{4423} - 1$ is a prime—in fact, one of the largest known. Lest we dismiss this information as insignificant, let us consider that the number has 1331 digits (in base ten) and is approximately 10^{1320} times as large as the number of electrons, protons, and neutrons in the universe! (See ex. 2.1.12.)

Though some computers may be able to calculate in seconds what formerly took humans a lifetime, they cannot provide a way of looking at the world or a method of proof that is not fed to them. Lacking such input, the computer by itself cannot prove or disprove that a particular formula (like those of Fermat or Mersenne) generates an infinitude of primes. As we shall soon see, however, this shortcoming does not by any means make the computer an insignificant instrument.

Numerous other formulas *may* represent answers to our latest requirements for a prime-generating formula. Originally they were conceived as formulas that might generate *only* primes, that is, they were thought of as answers to the question we considered earlier.

Let us consider, for example, Euler's famous function

$$f(x) = x^2 + x + 41.$$

What is $f(x)$ if $x = 1, 2, 3,$ or 4? Substituting directly into the formula, we get the following values:

x	(x)
1	43
2	47
3	53
4	61

Notice that the first four values of $f(x)$ are prime. What do you get for $x = 5$, and $x = 6$? Is $f(x)$ prime in these cases? It turns out that for the first thirty-nine values of x, $f(x)$ is prime! Unhappily, the function breaks down at $x = 40$ and at $x = 41$. Thus, $f(41) = 41^2 + 41 + 41$, which is a composite. As a matter of fact, you should be able to figure out that $f(41)$ is a composite without actually calculating the value of $f(41)$. (See ex. 1.3.5(g).)

Are formulas that generate a long sequence of primes before breaking down simply anomalies? Surprisingly enough, it can be shown that for any desired n, we can find a formula that will generate n primes before breaking down. For example, a formula that generates 79 successive primes is $x^2 - 79x + 1601$. Like Euler's function, it unfortunately breaks down (for $x = 80$). Though these formulas seem promising, the question of whether or not they generate an infinite number of primes (even admitting some composites) is an open one. (It should not be hard to see that if a polynomial function—like Euler's, for example—admits one composite, it admits an infinite number of composites. See ex. 2.1.9.)

Using a computer, Stanislaw Ulam and others recently found that of all Euler numbers less than ten million, approximately 0.475 ... are primes (see Stein, Ulam, and Wells 1964). This is no *proof* that an infinite number of primes will be generated by such a formula, but it does furnish us with a probabilistic statement—one that we shall explore in section 3.2. As David Hawkins suggests, many simply stated problems in elementary number theory may be insoluble because they are among the "undecidable" mathematical propositions that Kurt Gödel proved must exist in any consistent arithmetical system (Hawkins 1958).

Computers may be incapable of proving or disproving such undecidable propositions, but they can, by virtue of their prodigious capacity for computation, adduce enough positive or negative instances of a statement culled from a huge number of "events" to encourage a probabilistic conjecture. In evaluating the role of the computer in helping us form such conjectures, we are reminded of the rather glib remark often made that computers can do nothing that humans cannot do themselves. R. W. Hamming points up the superficiality of such a remark by suggesting an analogy to modes of transportation:

> It is like the statement that, regarded solely as a form of transportation, modern automobiles and aeroplanes are no different than walking. . . . Many of us fly across the U.S. one or more times a year, once in a while we may drive, but how few of us ever seriously consider walking more than 3000 miles? The reason the statement is false, is that it ignores the order of magnitude changes between the three modes of transportation. [Hamming 1965, p. 1]

We return now to our search for other formulas that generate an infinite number of primes, allowing for the omission of some primes as well as the inclusion of some composites. A glance at a table of primes immediately suggests that many primes occur in pairs (often called twin primes). The first few are 3, 5; 5, 7; and 11, 13. It may be that for any prime p, the formula $p + 2$ meets our requirements. Certainly whoever first discovered this "formula" would never have raised his or her expectations to those of Mersenne, that is, believe that the formula would produce only primes, even if some are excluded. The reason for such quick pessimism would have been that $p + 2$ yields many obvious composites early in the game. For example, the second element of each of the following pairs is composite: 7, 9; 13, 15; 19, 21.

The question of whether or not there is an infinite number of twin primes is still an unanswered one. This is surprising, since the problem seems on the surface not very much more complicated than the one solved so elegantly by Euclid over twenty centuries ago.

So far we have not hit on a formula that definitely satisfies even the seemingly modest requirement of generating an infinite number of primes, though not necessarily all, while admitting any number of composites (perhaps even an infinite number). Perhaps by now you have discovered one or two such formulas on your own. The following two observations suggest several possibilities:

1. The number of primes is infinite;
2. With the exception of 2, all the primes in N are odd.

Two obvious and somewhat trivial formulas are n and $2n - 1$ for n belonging to N (recall ex. 1.2.6 (i)). An interesting question now is, How much can we generalize from these two formulas? That is, substituting natural numbers in order into each of these formulas generates an arithmetic progression—a sequence in which any two neighbors differ by the same constant as any other two neighbors. The fact that the set of all even numbers in N also forms an arithmetic progression is sufficient, however, to convince us that not all arithmetic progressions generate an infinitude of primes (see ex. 1.3.7). Which ones have a chance? We turn to that question in the following subsection.

On Arithmetic Progressions and the Infinitude of Primes

Let us review a little about arithmetic progressions. How do we represent them? In order to answer this, let us look at such progressions. The sequence 2, 5, 8, 11, 14, ... is an example of an arithmetic progression. The first term is 2. How do we represent each of the terms in the progression in terms of the first term, 2? The answer is easy: 2, $2 + \underline{3} = 5$, $2 + \underline{6} = 8$, $2 + \underline{9} = 11$, $2 + \underline{12} = 14$, ... Let us now look at each of the underlined terms. They are multiples of 3. That is, the first is $3 \cdot 1$, the second $3 \cdot 2$, the third $3 \cdot 3$, the fourth $3 \cdot 4$. Thus the progression could be expressed as follows: 2, 2 + 3·<u>1</u>, 2 + 3·<u>2</u>, 2 + 3·<u>3</u>, 2 + 3·<u>4</u>,.... Notice that 2 and 3 are "fixed values" for each term in the progression and that the underlined terms proceed through the set of natural numbers. If, instead of 2, we let a stand for the first term of the progression and d stand for the common difference, then a is the first term, $a + d \cdot 1$ the second, $a + d \cdot 2$ the third, $a + d \cdot 3$ the fourth, $a + d \cdot 4$ the fifth, and so on. In general then, the formula $a + d \cdot n$ generates all the terms of the sequence as n varies in the set N and as a and d remain fixed.

To get a feeling for which progressions have a chance of generating an infinite number of primes in N even if we allow the inclusion of many composites, let us list a few progressions together with the generating formula for each:

Progression	Generating Formula
(i) 2, 5, 8, 11, 14, 17,...	$2 + 3 \cdot n$
(ii) 12, 16, 20, 24, 28, 32,...	$12 + 4 \cdot n$
(iii) 2, 6, 10, 14, 18, 22,...	$2 + 4 \cdot n$
(iv) 6, 9, 12, 15, 18, 21,...	$6 + 3 \cdot n$
(v) 9, 15, 21, 27, 33, 39,...	$9 + 6 \cdot n$
(vi) 3, 8, 13, 18, 23, 28,...	$3 + 5 \cdot n$
(vii) 7, 13, 19, 25, 31, 37,...	$7 + 6 \cdot n$

Notice that the formula in each case can actually be thought of as a generating formula yielding all the terms in the sequence if we permit n to be 0 for the first term.

Which of the seven formulas might satisfy our criteria? In (i), though there are several composites (8 and 14), there also are a number of primes —2, 5, 11, 17. Therefore (i) seems a likely candidate. How about (ii)? What are the primes in N from this sequence? There are none. Therefore it looks as though (ii) is out of the running. (Try a few more terms to see if this is really so.) Similarly, (iii) does not seem to have a chance, for the only prime in N is 2. (Try a few more terms to see if you meet with better success with (iii) than we have so far.) Formulas (iv) and (v) also seem unlikely candidates, but (vi) and (vii) look more promising.

The potentially successful formulas, then, are (i), (vi), and (vii); the obvious failures seem to be (ii), (iii), (iv), and (v). What might account for the difference? Perhaps you have a number of conjectures at this point, and several may be on the right track whereas others may be correct only for the cases you have examined but break down when you try them out on randomly chosen progressions of your own. Before reading on, you might wish to play around with your hunches.

Let us now look at the likely candidates and compare them with the unlikely ones.

Possibly Succeed	Seem to Fail
$2 + 3 \cdot n$	$12 + 4 \cdot n$
$2 + 5 \cdot n$	$2 + 4 \cdot n$
$7 + 6 \cdot n$	$6 + 3 \cdot n$
	$9 + 6 \cdot n$

In each formula, look at the relationship between a and d. Notice that in the "possibly succeed" listing, a and d have no common factor other than 1. What happens in the "seem to fail" group? Do a and d have common factors other than 1 in each case? Obviously yes. Both 12 and 4 and 2 and 4 have factors of 2 and 4 in common, and both 6 and 3 and 9 and 6 have a factor of 3 in common.

You might now wish to review exercise 1.3.6. If a and d have a common factor greater than 1, then we can represent a as $t \cdot m$ and d as $k \cdot m$ for some t, m, and k. Then $a + d \cdot n = (t \cdot m) + (k \cdot m) \cdot n = m \cdot (t + k \cdot n)$. Why? Any number in the sequence must be divisible not only by itself and 1 but by m also. Thus no numbers (with the possible exception of the first one) in the sequence can be prime if a and d have a factor in common. Two numbers that have no factor in common other than 1 are said to be relatively prime.

We see, then, that if a and d have a factor in common for any arithmetic progression, it is impossible for the progression to possess an infinite number of primes (more precisely, the progression cannot have more than one prime of N). But what if a and d *are* relatively prime? Consider the progression generated by $3 + 4n$. The first few elements are 3, 7, 11, 15, 19, 23, 27, 31, 35. Six of the first nine elements are prime, but although all these elements are odd, and all primes in N are odd (except for 2), not all odd numbers are representable in the form $3 + 4n$. For example, 13 is not of the form $3 + 4n$. It is instead representable as $1 + 4n$ (where $n = 3$). All odd elements of n are representable either in the form $1 + 4n$ or $3 + 4n$ (see ex. 2.1.13).

It is a logical possibility that after a certain point all primes must be of the form $1 + 4n$ and not $3 + 4n$. That this is not so is demonstrated in exercise 2.1.15. The general question of whether or not all progressions of the form $a + dn$ (with a and d relatively prime) generate an infinitude of primes has been answered in the affirmative. This conjecture was proved by Dirichlet (1805–1859), and the proof is quite complicated, invoking ideas that are not obvious from the set of real and complex numbers. In a few specific cases (like $3 + 4n$), the proof is relatively easy. It was not until about a century after Dirichlet's proof that a proof of the theorem was discovered (by Selberg in 1949) that did not require the use of properties of the set of real or complex numbers.

It is often surprising to discover that concepts involving real numbers or even complex numbers can be used or are even necessary in order to answer questions that appear on the surface not to involve them at all. In a perhaps distorted use of common language, any problem in number theory —no matter how complicated—that does not require the use of complex numbers in its solution is called *elementary*. In section 3.1 we shall discuss a theorem, first proved over eighty years ago, which was labeled nonelementary until a solution other than one using a complex number was found some thirty years ago.

Summary

We have proved in section 2.1 that there are an infinite number of primes. In addition we have shown that unsuspected difficulties are encountered in an effort to find a formula that would generate *all* these primes. Even the modest request for a formula that would generate an infinite number of primes, though not necessarily all, was found to be more ambitious than suspected. After demonstrating the failure of the conjectures of Mersenne and Fermat ($2^{(2^n)} + 1$ and $2^p - 1$), we have claimed that this search was rewarded but in a rather peculiar way. That is, there is a formula that generates only primes, but our claim is an existential rather than a constructive one. Thus we know that there must be some number A having the property that $[A^{3^n}]$ generates an infinite number of primes, but we have no idea of the nature of A itself. Answers like that surely suggest that the gods are playing tricks on us.

In desperation we relaxed our demands once more and asked whether there are formulas that generate an infinite number of primes even if we allow for many (perhaps an infinite number) omissions as well as the inclusion of many (again, perhaps an infinite number) composites. The twin-prime conjecture (that there are an infinite number of primes that differ by two) as well as the formulas of Euler, Mersenne, and Fermat may fit into this category, but we have no proof of this. Some very mild reflection suggests that this last is our most promising line of inquiry. The formula that generates all the odd numbers of N does work, since there are an infinite number of primes, and beyond 2 all primes must be odd.

We have also indicated (and shown that it is *not* obvious) that many arithmetic progressions satisfy our most relaxed requirements for a primegenerating formula, though we have demonstrated how absurd it would be to expect all arithmetic progressions (those for which a and d are not relatively prime) to do so.

PROBLEM SET 2.1

1. Calculate the first few values of K_i in Euclid's proof of the infinitude of primes:

$K_1 = 2 + 1 = 3; K_2 = 2 \cdot 3 + 1 = 7; K_3 = 2 \cdot 3 \cdot 5 + 1 = 31.$

A reasonable conjecture at this point might be (despite our claim in the text that Euclid's proof does not settle this question one way or the other) that K_is are always prime. Calculate K_4 and K_5. Are these numbers prime? How does this affect your belief in the conjecture? Try K_6. What is 30 031 ÷ 59? Now examine once more our original conjecture. How would you modify it? Now take another look at Euclid's proof of the infinitude of primes.

2. Exercises 2 and 3 are prerequisites for exercise 4. Look at exercise 4 before backtracking to exercises 2 and 3. If exercise 4 looks overwhelming to you now, you might wish to pass it by and return to it at a later stage. It is not essential for future understanding except for parts of chapter 4. For each of the following subsets of N, what is the *least* (or smallest) element (if it exists)?

a) N itself
b) $\{1, 2, 3, 4\}$
c) $\{1, 7, 50, 93, 101\}$
d) $\{x: 1 < x < 10\}$. The notation "$x:$" means "all x such that."

3. For each of the following subsets of the set of positive rational numbers (fractions like 1/3, 2/7, 8/1, 27/3, etc.), what is the least element (if it exists)?

a) The set of positive rationals itself
b) $\{1/2, 1/4, 1/3\}$
c) $\{x: 1 < x < 10\}$
d) $\{x: x^2 > 2\}$

4. Exercises 2 and 3 tend to point out an almost obvious principle that the set N obeys but that the set of rationals does not. The principle is called the *well-ordering property* and we shall come to it again in chapter 4. It asserts that any subset of N has a least element. This property is very useful in proving that *any composite in N must be divisible by a prime*, a conclusion we used in Euclid's proof that there are an infinitude of primes. Here is the proof. Can you supply the missing reasons?

Proof (by contradiction). Suppose there is a number in N having the property that it is composite but not divisible by a prime. There must be a smallest such number (why?). Let n be the smallest number. If n is composite, it has some factor other than 1 and n. Call that factor b. Thus $n = b \cdot a$ for some a, b belonging to N. We know from our experience (and this we could make more precise, as we do in chapter 4) that $1 < a < n$ and $1 < b < n$. Since a and b are less than n and since n is the smallest composite not divisible by a prime, a and b are both divisible by some prime. Call the primes p_a and p_b, respectively. Then $a = p_a \cdot x$ and $b = p_b \cdot y$ (why?), and $n = (p_b \cdot y) \cdot (p_a \cdot x)$. But then p_b divides n, which contradicts the hypothesis that there exists a composite in N that is not divisible by a prime, and our proof is complete.

5. Look again at the conjectures of Fermat and Mersenne. Do you think the following revised formulas have a chance of succeeding in the sense in which Fermat and Mersenne had hoped theirs would succeed?
  a) $F'(n) = 2^{2^n} + 2$
  b) $M'(p) = 2^p - 2$ for p a prime

6. If Mersenne's conjecture for generating primes were correct, then we would actually have a very easy and constructive method for demonstrating the infinitude of primes. All we need to know is that one prime exists, for example, 2. Then, if we let $q_1 = 2$, his formula would tell us that $q_2 = 2^2 - 1 = 3$ is prime. Applying his formula once more (using the prime q_2 in the formula), we would be able to conclude that $q_3 = 2^{q_2} - 1 = 2^3 - 1 = 7$ is prime. What would q_4 be? Given any prime q_n in this sequence, how would we generate q_{n+1}? What is the first value of n for which you have some difficulty determining whether q_n is in fact prime or not?

7. In discussing the work of Mills, we introduced the idea of the "largest integer not exceeding x," denoted by $[x]$. Thus,
$$[9] = 9, \ [8 + 2/3] = 8, \ [8.999] = 8.$$
  a) Answer the following questions (if possible without actually calculating).
    (i) $\left[\dfrac{3.99999}{2}\right] =$
    (ii) $[(2 + 1/2) + (3 + 1/3)] =$
  b) For the following, try several examples and indicate whether you think the generalization is true or false, and why:
    (i) For all x, y, $[x] + [y] = [x + y]$
    (ii) For all x, $[x] = x$

8. With regard to Euler's function $f(x) = x^2 + x + 41$—
  a) Complete the following table by substituting elements of N for x:

x	$f(x)$
1	43
2	47
3	53
4	61
5	?
6	?
7	?
8	?

b) Using the following hint, how might you have figured out the four entries above without such laborious calculation?

x	$f(x)$
1	43
	) 4
2	47
	) 6
3	53
	) 8
4	61
	) 10
5	?
	) ?
6	?
	) ?
7	?
	) ?
8	?

c) Can you find a number other than 40 and 41 for which Euler's formula fails to yield a prime? (*Hint*: Try 2×41.)

9. Show without actually calculating directly that 123 (which equals 3×41) also yields a composite when plugged into Euler's function. Find three other values for x that yield a composite. Show that there are an infinite number of "failures."

10. With regard to the function $f(x) = x^2 + x + 2$—

a) Make a table as in 8(a). Do you think you will generate any primes of N with such a formula? Why or why not?

b) Look once more at 8(b). Can you use a similar procedure with $x^2 + x + 2$ to fill in the table? Try it. Choose your own quadratic function (a function of the form $ax^2 + bx + c$), specifying a, b, and c as you wish, and see if the shortcut implied in 8(b) works.

11. Consider the function $f(x) = x^2 + x + 5$. How many primes does it generate before breaking down?

12. We claimed in this section that $2^{4423} - 1$ has 1331 digits in base 10. You can get some feeling for how we arrived at this conclusion by filling in the following table:

Number	Power of Ten	Number of Digits
$\overline{10}$	10^1	2
$\overline{100}$	10^2	3
$\overline{1000}$	10^3	4
$\underbrace{1000,\ldots,0}_{n \text{ digits}}$	$10^?$	?

Consider numbers that are not integral powers of 10, for example, 123. It follows that

$$10^2 < 123 < 10^3.$$

Try several other examples to convince yourself that for x any n-digit number,

$$10^{n-1} \le x < 10^n.$$

Then x may be expressed in the following form: $10^{(n-1)+\text{fraction}}$ where the fraction is between 0 and 1. In this case, we can calculate that 123 equals approximately $10^{2+0.09} = 10^{2.09}$. This analysis suggests that if $x = 10^n$ for some number n (not necessarily an integer), then the number of digits in x is $[n] + 1$. If n is a nonnegative integer, then—as the table you filled in suggests—"[ ]" may be eliminated, since $[x] = x$ only when x is an integer. (See ex. 2.1.7(b)ii.)

To figure out the number of digits in $2^{4423} - 1$, we need to appreciate that the number of digits in $2^{4423} - 1$ is the same as the number in 2^{4423}. Thus to get the number of digits in $2^{4423} - 1$, solve the following equation:

$$2^{4423} = 10^n$$

What, then, is the number of digits in terms of n? If the number of digits is 1331, what is $[n]$?

13. We showed that not all odd elements of N are representable in the form $3 + 4n$. Why? Find four odd elements that are not representable this way. We claimed that all odds in N are either of the form $1 + 4n$ or $3 + 4n$ for n belonging to N (allowing for the possibility that $n = 0$. Why?). To get some insight into what is going on, divide all elements of the form $3 + 4n$ by 4. Why do you believe that *all* odd elements must be of one or the other two forms already described? Show that other variations (like $5 + 4n$ or $7 + 4n$) are really of the two forms we have already considered, though in a slightly disguised state.

14. Prove that the product of two numbers of the form $4n + 1$ is of the same form. That is, show that $(4j + 1) \cdot (4k + 1) = 4(?) + 1$. What does (?) equal in terms of j and k? As a start, perhaps it would help to let $4j = A$ and $4k = B$. How, then, can you express $(A + 1) \cdot (B + 1)$ as a sum rather than a product? If you are stuck, look at the problem geometrically, as shown here:

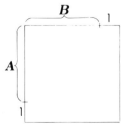

How can you break up the regions in the large rectangle into smaller regions? What are their areas? If it helps, try the same problem with Cuisenaire rods. Choose two of the smallest blocks for your unit, and select any other rods for A and B. Make a rectangular region, as above, and see what smaller rectangular regions can compose the larger ones. What are the areas of these regions in terms of the unit rod?

15. This exercise is a little tricky. It requires a proof by contradiction. Before attempting it, try exercise 13 and 14 and review Euclid's proof of the infinitude of primes. Our object is to prove that there are an infinite number of primes in N of the form $4n + 3$. We begin, as Euclid did, by assuming that there are at *most* n primes of the desired form

$$q_1, q_2, q_3, \ldots, q_n.$$

We now construct a new number that is analogous to Euclid's K_is. Choose $L = 4 \cdot (q_1 \cdot q_2 \cdot \ldots \cdot q_n) + 3$. How does L differ from Euclid's K_n? Then is L prime or composite? As with Euclid's proof, let us analyze both possibilities:

(i) If L is prime, we are done. Why?

(ii) If L is composite, then it must be divisible by at least one prime. The primes $q_1, q_2, \ldots, q_n$ will not work, since in all cases we are left with a remainder of 3 (the only number that will not give such a remainder is 3 itself, but we can eliminate the possibility that one of the q_ns is 3, since 3 is not expressible in the form $4 \cdot n + 3$ for $n \in N$.) Therefore there must be some other prime that divides L. The proof is not finished, however, for we must still demonstrate that the new prime is of the form $4 \cdot n + 3$ and not of the form $4n + 1$. Let us proceed by contradiction. Suppose there were *no* primes of the form $4n + 3$ that divided L. Since L is odd and cannot be divisible by an even number (ex. 1.2.5), it must be divisible only by primes of the form $4n + 1$. But the products of primes (or of any numbers, for that matter) of the form $4n + 1$ are of the same form (ex. 2.1.14). Therefore there must be at least one prime other than q_1 through q_n (for reasons we have already mentioned) of the form $4 \cdot n + 3$ that divides L (for otherwise L could not be expressible in that form).

Go through this proof several times and compare it with Euclid's proof of the infinitude of primes. Is anything (concerning primality) being claimed with regard to L? Answer the question specifically in light of exercise 2.1.1.

16. If you understood exercise 15, try the same scheme to prove that there are an infinite number of primes of the form $4 \cdot n + 1$. Compare and contrast the two proofs.

17. We have claimed, in Euclid's proof of the infinitude of primes, that if K_n is composite, then there must exist some new prime that divides it. We said that the proof *alone* is a nonconstructive one, in that it does not tell us how to find that new prime. This suggests the possibility that we *can* construct the new prime but that the proof alone does not tell us how. Discuss this possibility. Try to find someone who disagrees with you, and return to the debate again after reading section 3.1.

2.2: IN *E*

Thus far we have reviewed some interesting problems related to the number and generation of primes in *N*. With section 2.1 as a reference, we are now prepared to ask similar questions dealing with the domain *E*. Before continuing with the analysis in this section, you might find it profitable to analyze analogous problems in *E*. (It might be helpful to review the summary in section 2.1.)

Reconsider the first question posed in *N* from our new perspective. How many primes are there in *E*? We have shown (section 1.4) that the first few are 2, 6, and 10. What is the next prime in *E*? The next possible candidate, 12, will certainly not do because it is divisible not only by 12 and 1 but by 2 and 6 as well. (Why is it not divisible by 4 or 3?)

What about 14? Is it prime in *E*? The only reasonable factors that might conceivably destroy its status as a prime are 7 and 2. Applying the definition of factor in section 2.1 will allay all fears, however: recall that 2 and 5/2 are *not* factors of 5 in *N*. Therefore, 14 is prime. So far, then, our list of primes in *E* is 2, 6, 10, 14.

Perhaps you are prepared at this stage to make a not-too-risky conjecture. What would you guess to be the next term in the sequence? Persuade yourself that you are right, and show that 18 is the next term in the sequence.

Let us now look again at the sequence of the first five primes in *E*: 2, 6, 10, 14, 18. How many primes do you think there are in all? How would you describe this sequence? One's first observation might be that they differ by 4. Notice that the composites differ by 4 as well and that the sequence for the first few of them would be 4, 8, 12, 16. A reasonable way of distinguishing the composites from the primes in *E* might be to conjecture that the primes begin with 2 *and* differ by 4. This is true, but can you find a more explicit formula that would generate all the primes in *E*?

As a consequence of the definition of prime, every prime *p* in *E* must have two different factors in *E*. (What are they?) So 14, for example, has two factors in *E*, 1 and 14. If we consider the factors of 14 in *N*, however, then 14 is divisible by 2 and 7. Similarly, 10 is divisible by 2 and 5 if we focus on *N* rather than *E*; 6, by 2 and 3. Notice that every prime in *E* seems to be representable as twice an odd element of N. If this observation can be generalized, then we realize that we have an infinite number of primes in E, since there are an infinite number of odd elements in N.

Why should two times every odd element of N be a prime in E? Let θ be an odd element of N. Then we are saying that $2 \cdot \theta$ must be prime in E. That is to say, the only factors in E of $2 \cdot \theta$ are 1 and $2 \cdot \theta$. What else could possibly work? θ itself certainly does not divide $2 \cdot \theta$ in E, because θ does not belong to E. Similarly, 2 does not divide $2 \cdot \theta$ in E, because there is no element of E such that 2 times that element equals $2 \cdot \theta$.

The observations above, though necessary, do not yet settle the fact that there are no factors of $2 \cdot \theta$ in E other than 1 and $2 \cdot \theta$. Though θ certainly does not divide $2 \cdot \theta$ in E, how do we know that there is not some element of E smaller than θ that divides $2 \cdot \theta$? How do we know whether or not there is some element of E *between* 1 and $2 \cdot \theta$ that divides $2 \cdot \theta$ even if it does not divide 2 or θ? For example, in N there is some number between 1 and 12 that divides $4 \cdot 3$ and yet does not divide 4 or 3. (What is it?)

The questions above are presented to get you to reflect a little on a point that may be obvious on the surface but in fact has some substance to it. If you are not sure how to attack these questions, you might find it helpful to review exercises 1.2.6v and 2.1.4. We shall suspend our discussion of formal methods of analyzing these questions at this point but continue it in section 3.3.

You should persuade yourself in some way that it is reasonable to believe that twice an odd element of N is always prime in E and therefore that there are an infinite number of primes in E just as in N. As a first step in persuading yourself of this relationship between N and E, you ought to search for a counterexample—something that would disprove the assertion.

Notice that in this section we have merged the issues that we handled independently in section 2.1. In section 2.1 we asked, first, how many primes there were, and second, whether we could find a formula to generate them. We see that in E there is a much more intimate connection between the two aspects of the problem. The fact that twice an odd element of N always yields a prime in E informs us of two things almost simultaneously: (1) that there is an easy formula for generating primes, and (2) that there also must be an infinite number of primes. As a matter of fact, as we shall see in section 3.3, we essentially kill another bird (the determination of primality) with the same stone. It will turn out that one rather elementary idea in E provides the key to three or four sticky and seemingly independent issues in N. We now turn to a closer look at the generation of formulas in E for the purpose of drawing comparisons with N.

A Closer Look at the Generation of Primes in E

We have suggested that since twice an odd element of N always yields an element of N that is a prime in E, there must be a simple formula for generating primes in E. We have not yet stated such a formula explicitly, however.

Since all odd elements of N can be generated by the formula $2 \cdot n - 1$ (ex. 1.2.6i) for $n \in N$, the obvious formula that generates all primes of E is

(1) $\qquad 2 \cdot (2 \cdot n - 1)$, for all $n \in N$.

Though (1) *does* generate primes and only primes in E and is a perfectly respectable formula, from an aesthetic point of view it leaves something to be desired. Take another look at it. We are trying to generate primes in E, but what is the domain of the variable n? Notice that—

- by letting $n = 1$, we get $2 \cdot (2 \cdot n - 1) = 2$, the first prime in E;
- by letting $n = 2$, we get $2 \cdot (2 \cdot n - 1) = 6$, the second prime in E;
- by letting $n = 3$, we get $2 \cdot (2 \cdot n - 1) = 10$, the third prime in E; and so on.

We must substitute all the elements of N (including those that don't belong to E) into a formula in order to generate the primes of E. We must go "outside" E (or beyond E) to find a formula to generate primes in E. Is there a way of avoiding the use of elements of N? The situation is analogous to employing a formula with the domain Q of rational numbers to generate primes in the more restricted domain of N.

Since the attempts (unsuccessful as they were) at generating *only* primes in N did involve only the set N itself (as with Fermat, Mersenne, Euler), it would be desirable to discover a comparable formula whose domain is E (and not a more inclusive set than E) that would generate the primes in that set. Luckily a little bit of elementary manipulation leads us to a generating formula for primes in E that does not require recourse to the set N.

Let us start with our original generating formula $2 \cdot (2n - 1)$ and express it in an equivalent form that focuses only on the set E:

(2) $\qquad 2 \cdot (2n - 1) = 4 \cdot n - 2.$ (Why?)

We are almost done. Notice that $4 \cdot n$ for n belonging to N must be an even number (why?). Yet the variable n is still chosen from the set N and we have not yet formed an expression that involves only elements of E. Look again at $4 \cdot n - 2$. How would you rewrite it so that the values that get substituted into the formula are elements of E and all the primes in E are generated? The following should give a clue to what would be a reasonable approach:

(3) $\qquad 4n - 2 = 2 \cdot (2n) - 2$

We are really done now, for $2n$ must always be an element of E if $n \in N$. So if $n = 1$, then $2n = 2$; if $n = 2$, then $2n = 4$; if $n = 3$, $2n = 6$; if $n = 4$, $2n = 8$; .... Notice that the sequence formed by $2n$ is 2, 4, 6, 8, ... If

31

we let $2n = e$ for n any element belonging to N, we can see how to produce a formula for generating primes in E that uses only the domain of E:

(4) $$2 \cdot (2n) - 2 = 2e - 2$$

The right side of (4) is the formula we are looking for. That is, $2e - 2$ yields all (or almost all) the primes in E as we substitute in all the elements of E. Why did we say "almost all"? Substitute all the elements of E into $2e - 2$ and see if you get a prime of E every time. You will encounter a difficulty in the first substitution, since if $e = 1$, $2e - 2 = 2 \cdot 1 - 2 = 2 - 2 = 0$, and 0 is not a prime in E. Are there difficulties in further substitutions? Try $e = 2$, $e = 4$, $e = 6$, $e = 8$, ... Do you get a prime in E each time? Persuade yourself that this is so. Given the following obvious qualification, we thus have a formula for generating primes in E that does not require that we focus on N (a larger domain):

(5) $$2e - 2 \text{ for } e \in E' \text{ generates primes of } E.$$

On Revising the Generating Questions

Recall the three questions we posed in section 2.1 with regard to the generation of primes in N:

1. Is there a formula that yields all prime numbers?
2. Is there a formula that generates an infinite number of primes (and only primes) even if some are omitted?
3. Can we suggest a formula that will generate an infinite number of primes, allowing for—
    a) the omission of many primes from the sequence;
    b) the inclusion of many nonprimes as well?

It was difficulty in answering question (1) that must have led number theorists to retreat to question (2), and it was problems with question (2) that must have led to the still more modest question (3).

Look once again at the prime-generating formula we arrived at in the preceding subsection. We claimed that $2e - 2$ for $e \in E'$ generates primes of E. Does this formula generate some of the primes in E? Does it generate all the primes in E? Does it include composites in E as well? That is, are we faced with the same kind of difficulty in E as in N? Must we retreat to less demanding territory because of the difficulties we encounter in answering what seem to be rather straightforward questions?

It should not be difficult to persuade yourself that in fact the generating formula $2e - 2$ answers question (1) itself! Imagine how thrilled any number theorist would have been to come up with such a simple formula for generating all primes in N.

The production of such a simple formula for E in a sense takes the punch out of questions (2) and (3) in E. For the sake of drawing analogies with investigations in N and for setting the stage for some interesting problems in 3.3, let us return briefly to a few comparisons. These comparisons will also help us review the results of section 2.1.

Consider first the formulas of Fermat and Mersenne, $2^{(2^n)} + 1$ and $2^p - 1$, respectively (where $n \in N$ and p is a prime in N). Try substituting elements of E (or even N, for that matter) for n and p in these two formulas and see if you can arrive at primes in E. Let $n = 1, 2, 3$ in Fermat's formula. What kind of number is $2^{(2^n)}$ for $n = 1, 2, 3$? Without too much effort, you should be able to persuade yourself that $2^{(2^n)}$ is always even (see ex. 1.2.6iii). What about 2^p (regardless of whether p is prime in N or in E)? Again it should not be difficult to see that 2^p is always even. If $2^{(2^n)}$ and 2^p are always even, then what kind of number must $2^{(2^n)} + 1$ and $2^p - 1$ be? They are, of course, always odd. We would therefore be quite surprised if the formulas of Mersenne or Fermat generated *any* primes in E, much less an infinite number of them. Both formulas are thus unfruitful in E.

Once we appreciate the total failure of the conjectures of Mersenne and Fermat to generate any primes in E, where do we go? One possibility is to toss our scrap work into the wastebasket and to look elsewhere for excitement. Another possibility would be to salvage any portion of these insights that could lead to more fertile ground for exploration. There are many ways of trying to salvage the conjectures of Mersenne and Fermat in E. Try some of your own. You might, for example, wish to doctor up the power to which 2 is raised in Fermat's $2^\square + 1$. That is, you might try substituting something other than 2^n in the box. Or you might wish to modify the sign between $2^{(2^n)}$ and 1. Are any of these modifications helpful? Try similar modifications with Mersenne's conjecture.

One interesting and perhaps unexpectedly fruitful suggestion is to modify the last number in the formulas of both Fermat and Mersenne—that is, to try putting a different number in the place of 1 in the formula

$$2^{(2^n)} + 1.$$

What happens? What becomes of their conjectures in E if we consider $2^{(2^n)} + 2$ and $2^p - 2$? Let us try the first few substitutions in these formulas, choosing the domain N:

n	$2^{(2^n)} + 2$		p	$2^p - 2$
1	6		2	2
2	18		3	6
3	258		5	30
			7	126

It should not be too hard to determine that the elements on the right-hand side of the two tabulations above are all primes in E. You might be able to see why this is true for all substitutions for n and p (whether from N or E), though the issue will be clearer after you read section 3.3.

If the claim above is true, then which of the questions ((1), (2), or (3)) stated at the beginning of this subsection do the modified formulas of Mersenne and Fermat satisfy in E? Compare the results in E with those in N for these modified formulas (see ex. 2.1.5). Before continuing with other comparisons in N and E, you may wish to try other variations of the formulas of Mersenne and Fermat to determine what the effect would be in E.

What becomes of Euler's conjecture in E? Recall that he originally proposed that the quadratic expression $x^2 + x + 41$ yields an infinite number of primes in N (though many are left out); that is, he supposedly answered question (2) posed at the beginning of this subsection. We claimed in section 2.1 that the conjecture yields primes for the first thirty-nine substitutions from N but unfortunately breaks down for $n = 40$ and $n = 41$. Thus $f(41) = 41^2 + 41 + 41$ is obviously divisible by some number other than 1 and $41^2 + 41 + 41$ (why does it break down at $f(40)$ also?). What other number divides $f(41)$? Not only does his quadratic formula not satisfy (2), but we suggested that it misses the mark by quite a bit, since an infinite number of composites are yielded by the formula (see ex. 2.1.9). Whether or not an infinite number of primes are also generated is, as we have indicated, an open question.

After having examined the formulas of Mersenne and Fermat in E, you should find it easy to show why Euler's $x^2 + x + 41$ has no chance of succeeding as a generating formula for primes in E. Regardless of what number we substitute for x (from N or E), what kind of number is $x^2 + x + 41$? You will perhaps wish to look back at Euler's table in section 2.1. His formula answers none of the questions (1), (2), or (3) in E.

You might now wish to approach Euler's conjecture in E in the same spirit in which you approached the conjectures of Mersenne and Fermat. How might Euler be resurrected in E? Obviously $x^2 + x + 41$ fails, but can we doctor it up? Is there any quadratic expression (of the form $ax^2 + bx + c$) that works in E? Try several of your own. In exercise 2.2.7 we examine the following rather simple quadratic expressions: x^2; $x^2 + x$; $x^2 - 2$. Before turning to that exercise, you might wish to try such expressions to see which, if any, have a chance of succeeding and to determine the sense in which they have a chance of succeeding (as pertains to questions (1), (2), or (3)).

What becomes of the infinitude of twin primes conjecture in E? Recall that in N, the (as yet unproved) conjecture is that there are an infinite number of primes that differ by 2. That is, there are an infinite number of instances in which both p and $p + 2$ are prime. Suppose p is prime in E; what can we say of $p + 2$? Recall that we established $2e - 2$, for $e \in E'$, to be the generating formula for all primes in E. If p is some prime in E, then it must be expressible in the form $2e - 2$ for some $e \in E'$. What, then, must be the form of $p + 2$? Clearly $p + 2$ is expressible in the form $(2e - 2) + 2$, which equals $2e$. But what must the nature of $2e$ be in E? Recalling that to be prime a number must have exactly two different factors in the domain, we see that $2e$ is never prime in E. As a matter of fact, $2e$ is always divisible by numbers of E other than 1 and $2e$. That is, it is obviously divisible by 2 and e as well (recall that $e \in E'$). Therefore we can claim with assurance that if p is prime in E, $p + 2$ must be composite. Not only does the twin-prime conjecture in E fail, but it fails very badly: There are *no* twin primes in E.

What do we do now about the twin-prime conjecture in E? It appears that there is very little left to do. Is it possible to salvage the conjecture of an infinite number of twin primes in E in the sense in which we attempted to salvage the conjectures of Mersenne, Fermat, and Euler? We are in for an interesting surprise here, provided we allow ourselves to reinterpret the notion of twin primes in N itself! Since with one exception (which?), primes are sought from among only the odd elements of N, another possible interpretation of the twin-prime conjecture in N would be that given a prime p, we look for the next potential candidate. In N, since primes (with the one exception) are found only among the odd numbers, given a prime, the next potential candidate is the next odd number that is of the form $p + 2$ if p is prime.

Of course, the next potential candidate does not always get "elected," as we have shown in section 2.1. Nevertheless, it would be foolhardy even to search for a candidate that differs by less than two from any prime p in N.

What happens if we use the next-potential-candidate concept in searching for twin primes in E? You might reconsider the list of primes in E as a clue. They are 2, 6, 10, 14, 18, 22, 26, ... It looks as though the next potential candidate for primality, given that p is some prime, is of the form $p + 4$. Is $p + 4$ always a prime if p is a prime, or does it only work for the first few elements of E? Some very elementary reshuffling is revealing: if $2e - 2$ is a prime, p in E, then how do we express $p + 4$? Obviously $p + 4 = (2e - 2) + 4 = 2e + 2$. We know that any number of the form $2e - 2$ for $e \in E'$ is a prime. What about any number of the form $2e + 2$ for $e \in E'$? There are several ways you might persuade yourself that any number of the form $2e + 2$ is also prime. Interestingly enough, this formula just misses generating all primes in E if we systematically substitute into it all elements of E' (how?).

Regarding the twin-prime conjecture, it is worth stressing that in N we are left with a simple-looking conjecture that has eluded the grasp of competent mathematicians for centuries: it has been neither proved nor disproved by a counterexample. In E, though the conjecture lends itself to two possible interpretations, both of them are very easily established; in E it's "all" or "nothing"—a theme we shall see repeated. In chapter 5 we shall reexamine the question of twin primes from the point of view of parity (oddness and evenness). This will lend further weight to the "all" interpretation.

In concluding this section, let us turn to an examination of Dirichlet's problem in E. Recall that he proved in N that all arithmetic progressions of the form $a + dn$ generate an infinite number of primes, provided that a and d have no factor greater than 1 in common (that is, they are relatively prime) and that N is the domain of n.

What arithmetic progressions generate an infinite number of primes in E? In order to feel the force of the analogy and for the sake of elegance, let us consider the domain of a, d, and n to be E'. Is it necessary for a and d to be relatively prime for $a + d \cdot n$ to generate an infinite number of primes in E? Try several progressions of your own (for example, $2 + 8 \cdot n$; $2 + 6 \cdot n$; $4 + 8 \cdot n$; $4 + 6 \cdot n$; $6 + 12 \cdot n$) to get a feeling for what the answer might be.

Realizing that 2 and 8 are not relatively prime in E and looking at the first few terms of $2 + 8 \cdot n$, we can begin to appreciate what the answer is. The first few terms are 18, 34, 50, and 66, and they are all prime in E. At this stage perhaps you have a way of persuading yourself that all elements of the progression $2 + 8 \cdot n$ are in fact prime in E. We shall see clearly in section 3.3 why this is so (ex. 3.3.7). It is therefore not necessary that a and d be relatively prime for $a + d \cdot n$ to generate an infinite number of primes in E. Is it sufficient, however? That is, if a and d are relatively prime, will the arithmetic progression yield an infinite number of primes? Try some progressions of your own in which a and d are relatively prime in E.

Let us now look at $8 + 6 \cdot n$. We know that 8 and 6 are relatively prime in E. Does $8 + 6 \cdot n$ generate an infinite number of primes? The first few elements of the progression are 20, 32, 44, and 68. Some simple division should persuade you that none of these numbers is prime in E. As a matter of fact, none of the elements of the progression $8 + 6 \cdot n$ will ever be prime, regardless of the value of n selected from E'. Again, why this is so will be discussed in greater detail in section 3.3.

We see, then, that the relative primeness of a and d is not sufficient to guarantee that $a + d \cdot n$ will generate an infinite number of primes in E. Can we *sometimes* get and generate an infinite number of primes if a and d are relatively prime? Try $2 + 6 \cdot n$ and $6 + 8 \cdot n$. They both generate an infinite number of primes, and a and d are relatively prime.

Notice that $6 + 8 \cdot n$ generates an infinite number of primes but $8 + 6 \cdot n$ does not. What is going on? It looks as though the relative primeness of a and d is not a criterion in determining whether or not the arithmetic progression $a + d \cdot n$ will generate an infinite number of primes in E. What is at stake?

Notice that of the several progressions we have discussed so far, $2 + 8 \cdot n$, $2 + 6 \cdot n$, and $6 + 8 \cdot n$ generate an infinitude of primes, but $8 + 6 \cdot n$ fails. There are a number of reasonable conjectures, based on this small sample, that you might come up with to explain the success of some and the failure of others. Notice what kind of number a is in the examples that succeed and in those that fail. Does it seem to matter what kind of number d is? After a period of trial and error followed by conjecture and an attempt to determine what's going on, you should persuade yourself that an infinite number of primes will be generated by $a + d \cdot n$ in E if and only if a is prime in E—regardless of the relationship of d to a (see ex. 2.2.2).

You may at this point wish to figure out which of the three generating questions posed at the beginning of this section are answered by progressions of the form $a + d \cdot n$ in E where a is prime in E.

PROBLEM SET 2.2

1. If $2e + 2$ or $2e - 2$ generates primes in E for all $e \in E$, what simple formula do you think generates all the composites in E?

2. We claimed that $a + d \cdot n$ will generate primes in E if and only if a is a prime in E. (We added the additional stipulation that—for purposes of analogy—the domain of n should be E' also.) Let us demonstrate that if a is a prime, then $a + d \cdot n$ will also be a prime in E—regardless of whether d is a prime or not. There are two possibilities: (i) d is prime in E, and (ii) d is composite in E. We shall prove the theorem for the first case, and leave it to the reader to prove the second. (d cannot equal 1, since we are choosing a, d, and n from E' and not from E.)

(i) If a is prime and d is prime, then $a = 2e - 2$ and $d = 2 \cdot f - 2$ for e, f belonging to E'. Since n belongs to E' also, $a + d \cdot n = (2e - 2) + (2 \cdot f - 2) \cdot g = (2e - 2) + 2fg - 2 \cdot g = 2(e + f \cdot g - g) - 2$. Since E' is closed under addition and multiplication, $e + f \cdot g - g$ belongs to E', and the last expression of the equation above is expressible in the form $2 \cdot j - 2$ for some $j \in E'$. Therefore we conclude what?

(ii) For the reader.

3. Go through an argument similar to the one above for the instances in which a is composite, showing that the number $a + d \cdot n$ must be composite. (*Hint*: See ex. 2.2.5 (c).)

4. Explain why it is that in E 2 and 4 are not relatively prime, and yet 8 and 6 are. Under what circumstances will two numbers be relatively prime in E?

5. Consider the domain T.

a) List the first six primes in T.

b) How many primes are there?

c) In E, the formula $2e$ for $e \in E'$ generates all the composites.

　　(i) What formula generates all the composites in T if you use only elements of T to generate?

37

(ii) What formula generates all the composites in T if you allow all elements of N to generate?

(iii) Can you find a formula that generates all the primes in T?

d) Using the "potential candidate" definition of twin primes, list the first three sets of twin primes in T.

e) Prove (in the same spirit in which a proof was offered in E) that the number of twin primes is infinite.

6. The nonexistence of a formula to generate all primes in N also implies that there is no way to predict what the nth prime is other than by enumerating all the predecessors and counting them. Thus, though no one knows what the billionth prime is, we should all be convinced that it exists. In E, of course, the problem is easily solved. The nth prime in E is equal to $2(2n) - 2 = 4n - 2$. What is the second, the tenth, and the billionth prime in E?

7. Euler's conjecture that $x^2 + x + 41$ yields an infinite number of primes obviously fails in E. Why? Let us look at several other quadratic expressions to determine which, if any, might have a chance of succeeding (and for which set of criteria—questions (1), (2), or (3), described earlier—they succeed). We shall prove the reasonableness of the conjectures you make in this problem in exercise 3.3.6.

a) Consider the quadratic expression $x^2 + x$:

x	$x^2 + x$
1	2
2	6
3	12
4	20
5	30
6	42

(i) Find the next two terms in the table.

(ii) Notice that each of the entries on the right-hand side of the table belongs to E. Why? Show that regardless of what element of N we choose, $x^2 + x$ will always be even.

(iii) Find the factors in E of each of the entries.

(iv) Which of the entries in the table are prime in E and which are composite?

(v) Make a conjecture regarding the ability of the quadratic expression to generate primes in E. Do you think the formula satisfies question (1), (2), or (3)?

(vi) How would you modify the conjecture if we selected x only from E rather than from N?

b) Consider the quadratic expression x^2. Consider questions (i) through (vi) above with this new formula (omitting (ii)—why?).

c) Do the same with $x^2 - 2$.

8. We claimed in this section that any number of the form $2e + 2$ for $e \in E$ is prime in E. Below are sketches of two proofs of this conjecture. We shall be vague about the domain from which the variable is selected; try to clear the matter up yourself. In addition, consider the circumstances (that is, what domain for e) under which $2e + 2$ generates all the primes of E.

a) In the subsection entitled "A Closer Look at Generating Primes in E," we derived $2e - 2$ as a generating formula for primes in E by making use of the fact that $2 \cdot n - 1$ generates all elements of N. One way of seeing that the formula is correct is by appreciating that, roughly stated, *odd numbers* can be thought of as one less than even numbers. Another way of thinking about them, however, is to appreciate that they can be expressed as one more than even numbers. Therefore (using what domain for n?) another generating formula for odds in N would be $2n + 1$. A simple formula for generating primes would then be $2 \cdot (2n + 1) = 4n + 2$. How do we get from $4n + 2$ to $2e + 2$ as a generating formula for primes in E? What domain do we have to choose for e?

b) Another derivation of the generating formula $2e + 2$ that makes direct use of the already accepted formula $2e - 2$ is as follows:

Can we express a number of the form $2e + 2$ as some number of the form $2e' - 2$? We leave it to the reader to worry about the domain of e' when we are done, but it should be obvious that the domains for e in the generating formulas $2e + 2$ and $2e - 2$ are very close to being E itself.

9. As for $2e + 2$ and $2e - 2$, is there a quick, intuitive way (other than ex. 8) whereby you might persuade yourself that if one of them generates all the primes in E, then so must the other (again, with an appropriate choice of domain in each)?

10. Can you find a formula that answers question (3)—stated earlier in this section—in E?

3

The Determination and Distribution of Primes

3.1: IN *N*: DETERMINATION

WE INDICATED in chapter 2 that we have no simple formula to generate all the primes in *N*. Where does that leave us? Of the questions that still intrigue us, which can we answer? To set the stage for a reasonable analysis of that question, look carefully at each of the following questions:

A.
  1. Is 9 prime?
  2. Is 3599 prime?
  3. Is $2^{4423} - 1$ prime?

B.
  4. What is the first prime?
  5. What is the tenth prime?
  6. What is the twentieth prime?
  7. What is the billionth prime?

C.
  8. What are the first ten primes?
  9. What are the first twenty-five primes?
 10. What are the first billion primes?

D.
 11. How many primes are there between 1 and 10?
 12. How many primes are there between 1 and 25?
 13. How many primes are there between 1 and 1 000 000 000?

Some of the questions above are easy to answer without knowing much more than the definition of a prime. Questions 1, 4, 8, and 11 are such questions.

What about some of the harder ones? An answer (in the affirmative) to question 3 was given in the text. Can we find answers for questions 2, 7, 10, and 13, however? More important, how do the sets of questions above relate to each other? For example, do the answers for questions in set D seem to depend on answers to questions in set C? Which of the hard questions are answerable and which are not? Perhaps you are beginning to wonder, "What does the author mean? Are these questions answerable, and if so, in what sense?" The "in what sense" is an important question to be asking. Compare questions 1 through 13 with the following kinds of questions:

1'. What is a good man?
2'. What is the cause of lung cancer?
3'. What are the reasons for which Ms. Rosamond poisoned her ex-husband?

How do the hard questions about primes compare in answerability with questions like 1'–3'? In what sense are some of the questions about primes hard? All the questions about primes are clear in that we can understand their meaning without too much effort. This is not so, of course, with questions like 1'–3' above. Here we must ask more questions, like "What do you mean by *good*?" In addition, it is conceivable that even if we clarify meaning, these questions may remain unanswered, for we may never acquire the kinds of research tools necessary to solve such problems as the cause of cancer or what motivates humans to behave the way they do. Furthermore, it is conceivable that even if humans were to live forever, they would not be able to answer questions like 1'–3'.

Let us grant humans immortality (which—provided we do not destroy ourselves in a nuclear holocaust—they have in a sense, even if the individual is mortal) and look once more at questions 1–3 in set A.

Keep in mind the following question throughout: Is there some sort of simple procedure we could follow to answer each of these questions even if the procedure takes a long time (perhaps more than a lifetime) to carry out? Let us look first at questions 1–3, those described in set A.

Is n a Prime?

Which of the first three questions is hardest to answer? Easiest? It is easy to answer the first question in the negative, since in addition to 1 and 9, 3 divides 9. The number therefore has three factors and is not prime.

What about the second question? Is 3599 a prime? Here the problem is considerably more time-consuming because of the size of the number. How can we tell if it is prime or not? We asked this question in an earlier problem set (ex. 1.3.5(*c*)), and perhaps you have had time to play around with it. What have you found? Since we are using this number as a case study, it is worth reading on even if you believe you have already answered it.

The number 3599 is considerably larger than 9, and it appears on the surface that part of the difficulty in answering this question is a result of the size of the number. This is certainly part of the difficulty, but it is worth stressing that it is not merely size that is involved. For example, it is immediately obvious that 3598, 3600, 3602, 3604, 3636 are composite, despite the fact that these numbers are all roughly equal in magnitude to 3599 (see ex. 1.3.7). Furthermore, 3605 is also composite, since it must be divisible by 5. The only challenging numbers, then, are odd numbers not ending in 5. So in the vicinity of 3599, the challenging possibilities are

$$3597, 3599, 3601, 3603, 3607, 3609.$$

There are several techniques at our disposal for determining the primality of 3599. Our original definition of primality has the most obvious appeal. Recall that a number is prime if it has exactly two different factors. What are they for any prime p?

Using the definition of primality alone, we could begin to divide 3599 by all possible candidates in searching for divisors other than 1 and 3599. But what are the candidates? Intuition might tell us that the only reasonable choices would be numbers between 1 and 3599. This is certainly so, for it would seem foolish to try a number greater than 3599. But why is this true? Why are we so sure that there are no numbers *greater* than any particular integer that divide it? On the surface it would seem that only a madman would consider such a possibility. On the presumption that both madmen and geniuses share important qualities, think about why this is so obvious. The question again is, How do we know that no element of N is divisible by a number that is greater than itself? Find some way of persuading yourself that the proof depends on some important property of the integers that not all sets share (see ex. 3.1.1).

Before proceeding, let us briefly summarize our steps in trying to find factors of 3599 in N:

a) Apply the definition. At first it appears that we could find factors from anywhere in N.

b) Narrow down the search by observing that if b divides 3599, then b is between 1 and 3599.

So far the task is still monumental—especially if we must rely on human calculators. Who wants to divide 3599 by all the numbers between 1 and 3599? We record a previous observation that lessens our burden substantially:

c) No even number could divide 3599 (see ex. 1.2.6v). Neither could a number ending in 5 divide 3599 (see ex. 3.1.2).

We have so far limited the possible number of divisors to

$$1, 3, 7, 9, 11, 13, 17, 19, \ldots, 3591, 3593, 3597.$$

What do the three dots mean in the previous statement? By about how many numbers would you have to divide 3599 if you were to employ all the candidates implied by the restrictions suggested in (*b*) and (*c*) (see ex. 3.1.5)? The number of possibilities is still quite unmanageable. What would you do to narrow down the search?

Look once more at (*b*). Here we are told a reasonable range within which we might search for divisors of the number. Can you narrow down the range slightly? One possibility seems intuitively obvious. Would you expect to find a number greater than 2000 that divides 3599? Analogously, would you search for a number greater than 15 that divides 29? Of course, there is one number greater than 2000 that divides 3599 and one number greater than 15 that divides 29. What is it in each case? These numbers work only in a trivial sense, however. We are looking for some number x such that $1 < x$ and $x < 3599$ (or $1 < x < 3599$). Our experience with numbers in N suggests that we never find divisors (other than the number itself) of a number that are equal in value to more than half the number. Let us record this observation below:

d) If $a|b$ (*a* divides *b*) in N and if $a \neq b$, then $a \leq b/2$.

As with (*b*) itself, this property is less easy to prove than one might imagine. The proof depends on a property of N that is not shared by all domains. To convince yourself that (*d*) is not as simple an observation as appears on the surface, try exercise 3.1.6. To see why your intuition leads you to accept (*d*), see exercise 3.1.7.

Can you now estimate how many divisors we would have to test before being confident in asserting that 3599 is prime? Of course, all you need do is take half your answer to exercise 3.1.5.

Let us try to do even better than (*d*) in searching for a range of divisors of 3599. We know that we need not go beyond the number one-half of 3599, which is roughly 1800. At first it might seem that we cannot do better than narrow our range to candidates before the half-way point, as the following analogous example illustrates: $36 = 2 \cdot 18$. Therefore we conclude that 36 is not prime, but we observe that though one of the factors—2—is less than $1/2 \times 36$, the other factor—18—is exactly equal to it. It is therefore possible for one of the factors to be exactly equal to half the number we are examining.

Recall what we are searching for, however. In determining whether or not a number is prime or composite, we are not interested in listing all its factors. We are merely out to determine whether or not it has even one factor that is not equal to 1 or the number itself. Though 18 is half the original number, 2 is certainly quite a bit less than half the original number. Can we be guaranteed that if a number is composite and we systematically divide by numbers from 2 on, we shall hit any factor that is always within a better range than half the original number?

Let us look once more at 36 as an example of a number whose factors we are examining. We list all the factors of 36 in pairs from 1 on:

Factors of 36 (in pairs)	Reversal Point
(1, 36)	(9, 4)
(2, 18)	(12, 3)
(3, 12)	(18, 2)
(4, 9)	(36, 1)
(6, 6)	

In each instance, two numbers multiply together to give 36. Look at the smaller of the two factors in each set. The set of smaller factors from each of the sets is {1, 2, 3, 4, 6}. The largest number in this set is 6. Furthermore, notice that in the last four factorizations of 36 we merely reverse factors that have already appeared in the top four factorizations. Thus, if we systematically divide 36 by numbers from 1 on, we find that any factor greater than 6 (e.g., 9) has a partner that previously appeared and that was less than or equal to 6 (e.g., 4). That is, no matter what the factorization of 36 in N, at least one of the elements must be less than or equal to 6.

How does 6 compare with 36? There are many ways of relating them, but let us now focus on one that will be helpful in narrowing our search for factors that might reveal whether a number is prime or composite. Notice that $6 = \sqrt{36}$. Try finding the factors (in pairs) of several small numbers (less than 100) of your own choosing (try also ex. 3.1.9). Notice that for any paired factorization of a number, the smaller factor will not exceed $\sqrt{n}$. Let us articulate this observation mathematically:

e) If $a \cdot b = n$ for elements belonging to N, then either $a \leq \sqrt{n}$ or $b \leq \sqrt{n}$.

Thus, we are saying that it is impossible for both a and b to exceed $\sqrt{n}$. In essence, then, we need not search for factors of a number n beyond a number whose value is $\sqrt{n}$. You ought to examine (e) to see if it is a statement (like (a) and (d)) that seems to depend on a property of N that is not shared by other domains (see ex. 3.1.10 and 3.1.11).

Where are we now with regard to the status of 3599? We are attempting to find its divisors and have finally narrowed our search to numbers with the following qualifications: They must end with 1, 3, 7, or 9; and they must be less than 60 (for $\sqrt{3599}$ is about $\sqrt{3600}$, which equals what?). That narrows the search quite a bit, for we now need search for possibilities only among the following candidates:

$\alpha = \{3, 7, 9, 11, 13, 17, 19, 21, 23, 27, 29, 31, 33, 37, 39,$
$41, 43, 47, 49, 51, 53, 57, 59\}$

There are now only twenty-three possibilities. We could give a division problem to each person in a class in order to find out if any of these work. You might try this approach.

Of course, we might have inquired about a number considerably higher than 3599, in which case the task would have involved considerably more than twenty-three divisors. Can we do anything to narrow down the options? Because we are concerned with a more general case than that of 3599, let us explore some possibilities that we can employ under any circumstances; afterwards, we shall zero in on 3599.

If a number is composite, then what can we say about its factors? Clearly it must have at least two factors unequal to itself or 1. We can say more, however. In section 2.1, in which we reviewed Euclid's proof of the infinitude of primes, we claimed that if a number is composite, then not only must it be divisible by *some* number (other than itself or 1), but it must also be divisible by some number that is itself prime (see ex. 2.1.4).

Since we are trying to determine whether or not 3599 is prime, we need to search for possible divisors only among the primes in set α. If 3599 is composite, it must be divisible by some prime.

If we use the guidelines we established in (a)–(e) of this section, it is certainly an easy task to determine which of the possible divisors in set α is itself a prime. (For the largest number, 59, we do not have to try integers larger than $\sqrt{59}$. So the only possible numbers that might "destroy" divisibility of 59 are 3 and 7. Since neither of these works, 59 is prime.) With very little calculation, then, we limit our set of possible "destroyers" of the primality status of 3599 to

$$\alpha' = \{3, 7, 11, 13, 17, 19, 23, 29, 31, 37, 41, 43, 47, 53, 59\}.$$

There are thus only fifteen possibilities. Before reading on, you might wish to test each of these fifteen numbers to see whether or not **3599 is prime.** Let us state clearly the scheme that enabled us to narrow even further the possibile divisors of our number in an attempt to determine primality:

f) If b is composite in N, then b must have a factor smaller than b that is prime.

Summary of the General Scheme

At this point let us summarize the procedures (a)–(f) that enable us to determine whether or not a number is prime. Though we were motivated in our search by the specific instance of 3599, notice that there is nothing special about our rules that distinguishes that number from any other odd number not ending in 5. That is, we could easily replace 3599 by any potential prime in N (meaning that we eliminate the possibility of evens and multiples of 5), and arrive at rules (focusing beyond 3599 for (a)–(c)) that would still hold. Notice that as we progressed to rules (d)–(f), we no longer referred to 3599 but spoke instead of any element of N. To be clear about the generality of our search procedure, let us restate (a)–(f)

45

to cover all potential primes. We shall then try to compress these rules in order to devise a procedure that requires a relatively small amount of calculation:

a') If $x \in N$ and $y|x$, then y belongs somewhere in N.
b') If $x \in N$ and $y|x$, then y is between 1 and x.
c') If $x \in N$ is odd and not a multiple of 5, then the only divisors of x end in 1, 3, 7, or 9.
d') If $y|x$ and $y \neq x$, then $y \leq x/2$.
e') If there exist any y and $z \in N$ such that $yz = x$ and $y \neq x$, then either $y \leq \sqrt{x}$ and $z > 1$ or $z \leq \sqrt{x}$ and $y > 1$.
f') If $x \in N$ is such that there is some number y ($\neq x$) that is prime and that divides x, then x is composite.

Notice that our rules are really of two types. Rules (*a'*), (*b'*), (*d'*), and (*e'*), the first type, all refer to divisibility in general in the set N. They narrow the search for divisors of an element of N, and rule (*e'*) represents the best limitation of range in searching for divisors of elements of N most of the time (see ex. 3.1.12 for a discussion of the qualifier, "most of the time").

We know, then, that to find all the divisors of any number n, a good rule to follow would be the following:

e'') Try dividing n by those elements of N between 1 and $\sqrt{n}$.

If we find just one divisor, we know that n must be composite.

Suggestions (*c'*) and (*f'*), which belong to the second type of rule, remind us that we are searching for potential primes and not just focusing on divisibility in general. Rule (*c'*) suggests that since it is only odd numbers that would possibly be prime in N (with the exception of 2), the only possible divisors must end in 1, 3, 7, 9. Rule (*f'*) suggests that if possible we look only at prime divisors, since we have established that if a number is composite, it must have at least one divisor (less than itself) that is prime (see ex. 3.1.13 for an opportunity to employ these suggestions).

Rules (*c'*) and (*f'*) could be summarized as follows:

f'') Try dividing n only by primes that precede n.

Rule (*e''*) combined with (*f''*) gives us a "workable procedure" for determining if any number is prime.

On the Meaning of "Workable"

We have developed an algorithm that would enable us to determine whether or not a given number is prime. We claimed that (*e''*) and (*f''*) give us a workable procedure. But what does this expression mean? You might wish to look once more at the introduction in section 3.1, where we asked, with regard to determining primality, "Is there some sort of procedure that could take a long time (perhaps more than a lifetime) to carry out?" We have, then, devised a clear procedure to determine whether or not any number is prime. Furthermore, notice that this procedure involves a lot more than the definition of prime. We have to know how division behaves in N and have suggested in the exercises how comparison with other domains (especially Q, in this case) helps pinpoint properties that we frequently assume in N, but we haven't been mindful of the big leaps that such assumptions require (for example, that we need not search for divisors of $n \in N$ beyond n itself).

Though we have a clear procedure, it should be stressed that this procedure could be time-consuming, especially for very large numbers. However, a number of procedures can be used to ease the burden of computation slightly as one tries to determine the primality of a number (see ex. 3.1.19, for example). Still, for large numbers the task could be an overwhelming one in practice. It is true, as we have claimed before, that computers are very helpful in enabling us to answer the question, Is n prime? However, computers do have their limits.

Though there are a number of helpful tools that enable us to determine primality for *some* numbers, the general problem of determining primality for any number n is not susceptible to quick analysis. The problems are practical more than they are theoretical. Though we have determined that certain very large numbers are prime, question A (3) at the beginning of this chapter asks about one with 1331 digits, one of the largest known primes. We therefore have answers for only a very small number of elements of N. Tables of known primes do not exist for numbers much above one million, though computers are enabling us to extend this range considerably. In section 3.2 we shall describe a procedure for devising such tables.

Another point worth stressing is that it is not merely the size of numbers that poses practical problems for determining primality. For example, you can easily think of a number with one million digits that is not prime. What is it? You can think of one with a billion digits that is not prime (see ex. 3.1.14). It is the large numbers possessing a particular form that create the difficulty. That is, numbers that arouse our interest in terms of primality are large ones that end in 1, 3, 7, or 9 (assuming base-ten notation).

Furthermore, it is sometimes true that if a number $n \in N$ is composite (even among those that arouse our interest), then we may be able to determine this by some means other than the rather tedious one of dividing by all primes in N between 3 and $\sqrt{n}$. If n is prime, there is no such royal road. We of course are not guaranteed that composites will be revealed as such by nontedious means, but there is the possibility. Let us look at one illustration. Other examples are in the problem set.

Nontedious Approaches in a Special Case

What conclusion have you come to with regard to the primality of 3599? As we pointed out, there are only fifteen possible candidates once you have determined which numbers less than 60 are prime, and you should have the answer to the question by now. Let us look at the problem from a less tedious point of view, however.

The number 3599 is close to another number that in many ways is much more manageable. What number is it almost equal to? Motivated by this observation, we state

$$3599 = 3600 - 1.$$

Well, what's so appealing about $3600 - 1$? The advantage of 3600 over 3599 is that 3600 is the square of an element of N. That is, $3600 = (60)^2$. So,

$$3599 = (60)^2 - 1.$$

You probably know something about factoring such expressions. We know how to factor certain trinomials (e.g., $x^2 - 5x + 6$) and certain binomials (e.g., $x^2 - 9$). What about $(60)^2 - 1$? Can we factor it? If so, what is the advantage? $(60)^2 - 1$ is easily factored as the difference of two squares, once we have observed that 1 is also a square. Therefore, $3599 = 3600 - 1 = (60)^2 - (1)^2$. (See ex. 3.1.16 if you have a question about this.)

Further,

$$3599 = 3600 - 1 = (60)^2 - (1)^2 = (60 + 1) \cdot (60 - 1) = 61 \cdot 59.$$

Is 3599 prime or composite? Notice that if we do this problem by using the systematic procedures combining (e'') and (f''), then although we need examine only fifteen possibilities, we do have to go down to the wire, since the first prime that works is 59. Can you think of another nontedious method for demonstrating that 3599 is prime? (Try ex. 3.1.18.)

In addition to significant shortcuts of this sort in very special cases, we can in general use less dramatic devices to ease the burden of dividing one number by others. There are shortcuts, for example, to divide a number by 3 or 7. These are discussed in exercise 3.1.19. Thus the particular algorithm for division may sometimes be replaced by less demanding procedures.

One-Shot Trial Procedures for All Numbers

We have discussed clever one-shot procedures to determine primality that work for some numbers, for example, 3599 and 2 976 532 212 573-124 852. Is there no general one-shot procedure that works for all numbers? Given the absence of hand-held calculators, how many years of trying to calculate primes would you guess had passed before that question was raised? It must have been a particularly pressing question about three thousand years ago, before Euclid proved that there are an infinite number of primes (see section 2.1). Mathematicians must have waited with bated breath to see whether the largest known prime number would be topped.

In asking for a one-shot trial, we are actually asking for a procedure that would (1) minimize the trial-and-error character of the approach and (2) work for all numbers. There actually is such a simple procedure, one that requires a single division only.

To understand the procedure it is necessary to understand the term *factorial*, denoted by the symbol "!" in mathematics.

4! is shorthand for $4 \cdot 3 \cdot 2 \cdot 1$;

5! is shorthand for $5 \cdot 4 \cdot 3 \cdot 2 \cdot 1$;

7! is shorthand for $7 \cdot 6 \cdot 5 \cdot 4 \cdot 3 \cdot 2 \cdot 1$;

n! is shorthand for $n \cdot (n-1) \cdot (n-2) \cdot \ldots \cdot 1$.

First, calculate the value of n! for 4, 5, and 7 as above. Then do it for $n = 6, 8, 9, 10$. How much work seems to be involved?

In about 1770, two mathematicians arrived independently at the same procedure for determining primality: John Wilson (1741–1793) and J. L. Lagrange (1736–1813). It is actually a very simple procedure. Look at table 2 and see if you can guess what is involved.

TABLE 2

The First Five Instances of Wilson's Theorem

n	$(n-1)! + 1$	Does n Divide $(n-1)! + 1$?	Is n Prime?
2	$1 + 1 = 2$	yes	yes
3	$2 + 1 = 3$	yes	yes
4	$6 + 1 = 7$	no	no
5	$24 + 1 = 25$	yes	yes
6	$120 + 1 = 121$	no	no

Look at n and the answers in the right-hand column for each n. We have

2—yes
3—yes
4—no
5—yes
6—no

Does something look interesting here? Wilson's theorem is stated as follows: *For all n greater than* 1 *belonging to N, n is prime if and only if n divides* $(n-1)! + 1$. (See ex. 3.1.22–3.1.24 for a discussion of why the expression "if and only if" appears above and elsewhere in the book.)

Wilson's theorem is very pretty, but we get it at a price. If you tried it for 8, you have some idea of what is involved. It is easy to find out if 8 is prime. The numbers we divide by and the quotients we get all involve easy calculation. If we use Wilson's theorem to test for primality of 8, we need to find $7! + 1$. $7! = 7 \times 6 \times 5 \times 4 \times 3 \times 2 \times 1 = 5040$, and $7! + 1 = 5041$.

To see if 8 is prime, we divide it into 5041, and 8 is a very small number. You can imagine that although every question of primality is theoretically answerable by Wilson's one-shot divisibility test, the actual division quickly becomes overwhelming (see ex. 3.1.26 and 3.1.27). You might look back at the discussion at the beginning of section 3.1 to figure out in what sense Wilson's procedure is "workable."

PROBLEM SET 3.1

1. Our first labor-saving device, which aids us in determining whether or not a number in N is prime, depends on the observation that no factors of a number are larger than the number itself. Show that this property is not of the "any fool can see" variety by demonstrating that it is false in Q (the set of fractions). What property of N that is false in Q do you think is required in the proof of the statement? (*Hint*: Look back at ex. 2.1.4.)

2. Show that if $5|n$ for any $n \in N$, then n ends in 0 or 5 (assuming that n is expressed in base ten). Let us look at the proof for a number between 1 and 100. Any such number is expressible in the form $10t + u$ (for example, the notation 53 is shorthand for $10 \cdot 5 + 3$), where t and u may range between 0 and 9. If $5|10t + u$, show that $u = 0$ or 5. The proof is simple and depends on the following observations:

   a) $5|10t$. Why? Refer to the definition of *factor* (or *divides*) in section 1.4.

   b) If $5|10t + u$, then $5|u$. Here we rely on the more general observation that if $5|A + B$ and $5|A$, then $5|B$. Prove it by once more using the definition of *divides*. (Does this theorem depend on the choice of 5 as a divisor?)

   The proof might begin as follows:
   If $5|A + B$, then there is a C such that $5 \cdot C = A + B$;
   if $5|A$, then there is a D such that $5 \cdot D = A$;
   to show $5|B$, we need some number x such that $5 \cdot x = B$ (why?);
   we can represent B as $5C - 5D$.
   Now how can you represent the x so that $5 \cdot x = B$?

   c) Why must u (a number between 0 and 9) end in 0 or 5 for 5 to divide it?

3. Replace 5 by any N in exercise 3.1.2(b). Is the statement now true? That is, is the following true: If $N|A + B$ and $N|A$, then $N|B$? Try to show why this is true by using marbles or Cuisenaire rods in addition to an abstract proof.

4. We claimed in the exercise above that if $5|A + B$ and $5|A$, then $5|B$. Is this theorem true if we replace $+$ by $\cdot$? That is, can we say that if

$5|A \cdot B$ and $5|A$, then $5|B$? In exercise 1.2.2 and exercise 1.2.3 we demonstrated that if $2|A$ and $2|B$, then $2|A + B$. The proof for divisibility by 3 is equally simple. The general statement would be, If $C|A$ and $C|B$, then $C|A + B$. Is it true? Look back at exercise 1.2.3. How does this statement compare with exercise 3.1.3? Does the truth of one suggest the truth of the other? That is, could you prove exercise 3.1.3 by using the general statement above?

5. Using only restrictions (b) and (c) from the text (regarding the search for divisors of 3599), estimate how many candidates there might be. That is, roughly how many of the numbers between 1 and 3599 end in 1, 3, 7, or 9?

6. Given the set Q (of rational numbers) under the operation of multiplication, is (d) (in the text) true? That is, can we conclude that if $x|y$ and $x \neq y$, then $x \leq y|2$? Try a few rational numbers. For example, let $y = 17/3$. Can you find some rational number greater than $1/2 \cdot 17/3$ that divides $17/3$ in Q? An example of one number that works is 5. Are there others?

7. Though the proof of (d) is not presented in the text, we can gain some intuitive feeling for why it is true. In N, if a number b is not prime, then it has at least two factors other than b and 1. The smallest possible value that one of the factors can have is 2. What can you say, then, about the other one? If the smallest factor is greater than 2, then what must the other be (with regard to half the value of the original number)? To get some feeling for what is going on, try the number 36. If we are searching for factors other than 1 and 36, we might first come across 2. But then 18 is the other factor, and neither factor is greater than $1/2 \cdot 36$. What if the first factor we come across is 3? What is the other then? Is it less than 18, which is $1/2 \cdot 36$? Why? By the way, you might enjoy examining why the proof is intuitive and not rigorous. For one thing, you ought to compare assertions in this exercise with the example in exercise 3.1.6. In particular, which statements would be false? Is it true, for example, that the smallest possible factor of a number in Q (other than 1 and the number itself) would be 2?

8. List the total number of factors of 36 in N. How many are there? Is the number of factors even or odd? Answer the same set of questions for 49, 40, 32, 81, 18. How can you tell in advance whether the number of factors will be odd or even? You might now be interested in looking back at exercise 1.3.4.

9. List in pairs the factors for the numbers you chose in exercise 3.1.8. If you represent the number you start with by n, then what can you say about $\sqrt{n}$? When will there be a factor in a pair that is exactly equal to $\sqrt{n}$?

51

10. Is (*e*) of section 3.1 true for rational numbers too? That is, can we say that if a and $b \in Q$ so that $a \cdot b = n$ (for $n \in Q$), then $a \leq \sqrt{n}$ or $b \leq \sqrt{n}$? Try some numbers. For example, find any two fractions that multiply to give 4/9. What is $\sqrt{4/9}$? Is it possible for both of your factors to exceed $\sqrt{4/9}$?

11. One of the most straightforward ways of proving (*e*) of section 3.1 would be by the indirect method. That is, to show that—

e) if $a \cdot b = n$, then $a \leq \sqrt{n}$ or $b \leq \sqrt{n}$,

we could try to show that—

e*) if it is false that $a \leq \sqrt{n}$ or $b \leq \sqrt{n}$, then $a \cdot b$ cannot equal n.

Before proving (*e**), let us look at the logic involved. If we denote "$a \cdot b = n$" by "*P*" and "$a \leq \sqrt{n}$ or $b \leq \sqrt{n}$" by "*Q*," then (*e*) looks like the following: "If *P*, then *Q*." What does (*e**) look like? Denote "*P* is false" by "$\sim P$" (not *P*) and "*Q* is false" by "$\sim Q$" (not *Q*). We then wish to show, If $\sim Q$, then $\sim P$. Try to determine whether the statement "If *P*, then *Q*" is essentially the same statement as "If $\sim Q$, then $\sim P$." "Essentially the same" means that if one is true, then so must be the other, and if one is false, then so must be the other.

You have already been asked to employ this kind of reasoning in several places throughout this book. For example, in exercise 1.2.6vi, you were asked to show, by indirect proof, that if $n^2 \in N$ is even, then so is n. That is, if we denote "n^2 is even" by *P* and "*n* is even" by *Q*, then you were asked to show, If *P*, then *Q*. An indirect proof would involve your showing that if it is *not* so that *n* is even, then it is not so that n^2 is even. That is, you were required to show that if *n* is odd, then n^2 is odd.

Try nonmathematical examples as well as mathematical ones to determine whether "If *P*, then *Q*" is equivalent to "If $\sim Q$, then $\sim P$." You might enjoy reading sections on implication and truth tables in a book on logic to help you see what is going on.

Return now to (*e**). Show that if it is false that $a \leq \sqrt{n}$ or $b \leq \sqrt{n}$, then $a \cdot b \neq n$. When would it be false that $a \leq n$ or $b \leq n$? This occurs if $a > n$ and $b > n$. Suppose, then, that $a > n$ and $b > n$. Why is $a \cdot b > n$? We are essentially claiming that if r and t both exceed v, then $r \cdot t$ exceeds v^2. Persuade yourself that this is true. Look at positive and negative possibilities for r, t, and v. Is it true that $r \cdot t$ exceeds v^2 regardless of sign?

12. We claimed that (*e'*) was a better limitation of the range of divisors for any element of *N* than (*a'*), (*b'*), (*d'*) "most of the time." That's a peculiar kind of remark. In essence, what we are saying is that if a number *n* has divisors in *N*, we can find them all by dividing *n* by those elements of *N* between 1 and $\sqrt{n}$ at most. Of course, there may be factors of *n* that exceed $\sqrt{n}$. For example, 18 divides 36 yet it exceeds $\sqrt{36}$. The point,

however, is that we would have detected 18 as a divisor long before getting to $\sqrt{36}$ as a divisor by trying the number 2. At the very least we could have claimed that there was a nontrivial divisor (a divisor other than 1 or 36) of 36 whose value was less then or equal to $\sqrt{36}$, namely 2. (Even if we had found this out without discovering that its "partner" was 18, we would have been able to determine quite quickly that 36 was not prime.)

Let's return to the "most of the time" remark. We are claiming that the search for potential divisors between 1 and $\sqrt{n}$ (at least identifying one of the two partners whose product is n) is a narrower search than that for potential divisors between 1 and $n/2$. Is that true? What are we really claiming? We are claiming that $\sqrt{n} < n/2$ "most of the time." When is it false that $\sqrt{n} < n/2$? Try numbers from N in a systematic way. That is, try 1, 2, 3, 4, 5, ... Certainly, $\sqrt{9} < 9/2$, $\sqrt{16} < 16/2$, and so on. What about smaller numbers? Larger ones? Come up with an argument that is persuasive even if you do not believe it is airtight.

13. Using suggestions for testing for primality, estimate roughly the maximum number of divisors that would be required to determine whether or not the following are prime:

  a) 10 003   *c)* 101
  b) 1 000 001   *d)* 100 000 007

14. You can come up with at least one composite number with one million digits. One example is $10^{999\,999}$. How many other numbers with a million digits can you predict would be composite after only a brief inspection?

15. Comment on the following: If anyone claims (without recourse to a table of primes) to know whether or not n is a prime (for very "large" and "interesting" values of n), then you can be well assured that is *not* prime.

16. Prove that $x^2 - a^2 = (x - a) \cdot (x + a)$ both algebraically and geometrically.

a) Algebraically:

Show that $(x - a) \cdot (x + a) = x^2 - a^2$ by first considering $(x - a)$ to be a single term, y. Thus:

$$(x - a) \cdot (x + a) = y \cdot (x + a)$$

Now multiply y by $x + a$ (using what principle?):

$$y \cdot (x + a) = y \cdot x + y \cdot a$$

Now recall what y replaced:

$$y \cdot x + y \cdot a = (x - a) \cdot x + (x - a) \cdot a$$

Complete the calculation.

b) *Geometrically*:

Consider the following square within a square:

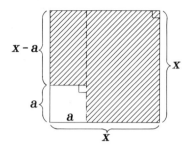

What we are trying to find is the shaded area. (Feel free to cut and paste.) If we do wish to express the shaded area as $(x - a) \cdot (x + a)$, then it is worth observing that we could do so by constructing a rectangle with $(x - a)$ as one of its dimensions. Do we already have a rectangle with a side $x - a$? Cut along the dotted line in the shaded area and rearrange the figure so that you have a large rectangle one of whose dimensions is $x - a$. What is the other dimension?

Can you find other ways of showing that $x^2 - a^2 = (x - a) \cdot (x + a)$ by using two squares as a starting point? What happens if you orient the two squares differently? Try putting the small square elsewhere in the large one.

What assumptions have we made about the relation of x to a? If the assumptions are false, do the geometric proofs still hold? Why or why not?

17. Show by nontedious means that 3551, 9991, and 22 499 are all composite. What perfect square is each of them almost equal to?

18. Demonstrate geometrically why 3599 is composite.

19. As you perhaps know, for some numbers you can find ways of easing the burden of divisibility by employing certain quick tests. In the following exercises we shall suggest tests for a few others. (More detailed justifications and examples appear frequently in the *Mathematics Teacher* and the *Arithmetic Teacher*, both published by the National Council of Teachers of Mathematics.)

a) *Divisibility by 3*. Let us create some large number that is divisible by 3 and some large number that is not divisible by 3. This is easy to do: for example, $3 \cdot x$ for any x in N must be divisible by 3. Why? Choose x to be 53 721, for example. (Try some values of x of your own choosing.) Then $3 \times 53\,721 = 161\,163$ is divisible by 3 and $161\,163 - 2 = 161\,161$, which is not divisible by 3. (Why not?)

Now do the following: Mix up the digits in each of the numbers any way you like. In the first instance, for example, change 161 163 to 161 136 or 161 361. Make *many* changes. What can you say about the divisibility of all these rearrangements with regard to 3?

Now do the same thing with 161 161. Make as many rearrangements as you can. For example, change 161 161 to 116 116. Try others. Are these new numbers divisible by 3?

From rearrangements of both these numbers, you should notice something interesting. If the first one is divisible by 3, it appears that all its rearrangements are too. If it is not divisible by 3, then none of its rearrangements are. Now what is the *constant* in all the rearrangements, either of the divisible or of the indivisible example? The position of the digits changes, but if you add up all the digits in each of the rearrangements for any number, what do you notice? Now you are ready to make a clear conjecture on how you can tell without dividing whether or not a number is divisible by 3.

b) *Divisibility by 5*. How do you know if a number is divisible by 5? This is an easy question. If you have trouble, reexamine exercise 3.1.2.

c) *Divisibility by 11*. Consider the number 9273. Is it divisible by 11? Try numbers that are created by rearranging every other digit of the original number. For example, consider 7293 (switching the first and third digits), 7392 (what have we switched here?), and 9372? Are these new numbers divisible by 11? Consider 9271 as a starting number. Is it divisible by 11? Now consider every other digit of this number, as we did before with the number 9273. Are these new numbers divisible by 11? Then see if you get the same result when you rearrange the digits as we did above. Do you have a conjecture? If you have trouble here, you might wish to reexamine part (*a*) of this exercise for a clue, though the procedure is more complicated here.

20. If you would like to prove why 19(*a*) works, you will find the following observation helpful. Look first at only three-digit numbers. Then we might express any such number as $100h + 10t + u$. Since our conjecture directs our attention to the *sum* of the digits, we must examine $(h + t + u)$. How can you express $100h + 10t + u$ in a new way so as to "isolate" $h + t + u$? One way would be the following: $100h + 10t + u = (99h + 9t) + (h + t + u)$. Now, using exercise 3.1(*e*) and whatever else you find helpful, complete the proof. Try to analyze 19(*c*)— divisibility by 11—by using a similar starting point.

21. The divisibility tests we have suggested both in the text of section 3.1 and in the exercises (see ex. 3.1.19, for example) assume that we have expressed our number in base ten. If you know something about bases other than ten, you might enjoy exploring what kinds of tests work in other

bases. For example, the numbers from one to ten are represented below in base two:

Base Ten	Base Two	Base Ten	Base Two
1	1	6	110
2	10	7	111
3	11	8	1000
4	100	9	1001
5	101	10	1010

a) Just by looking at the form of a number expressed in base two, how can you tell if it is even or odd? To what is this test analogous in base ten?

b) Write down a few more numbers expressed in base two. For base ten, we discussed a divisibility test for eleven. What would be analogous to eleven in base two? Does a test similar to divisibility by eleven work for that number expressed in base two?

c) Express numbers from one to ten in base three. Search for divisibility tests in that system. How do you tell if a number is even in that system?

22. We have claimed that n divides $(n-1)! + 1$ if and only if n is a prime. What is the function of the expression "if and only if"? In a sense it is the existence of "if and only if" that allows us to speak of Wilson's theorem as a test. Suppose we had claimed only that "n is prime *if* n divides $(n-1)! + 1$." We would then divide n into $(n-1)! + 1$, and if it worked, we would conclude that n is prime. But suppose it didn't work. That is, suppose n did not divide $(n-1)! + 1$. It is possible that n is *still* prime under these circumstances. That is, if all we know is that "n is prime if n divides $(n-1)! + 1$," then it is possible that n is also prime if n does not divide $(n-1)! + 1$. Then, of course, we would have no test for primality at all. Consider the following:

A friend tells you that he will not speak to you if you lie. Suppose you do not lie; what can you conclude? Suppose he tells you he will not speak to you if and only if you lie. If your friend keeps his word, what can you conclude he would do if you lie? If you tell the truth?

You may want to look back on a number of the tests for divisibility and see whether or not they have an "if and only if" character to them.

23. You know that in base ten a number greater than five ends in 1, 3, 7, or 9 if it is prime. Can we make that into an "if and only if" **statement**? Discuss.

24. What are some "if and only if" statements other than Wilson's theorem that you can make about the contents of this text?

25. Use Wilson's test to see if 11 is prime.

26. Is 23 prime or not? Try (but not too seriously) to answer the question with Wilson's procedure.

27. Use a hand-held calculator and try to find the largest number that you can check for primality with Wilson's test.

28. In section 2.1, we proved that there are an infinite number of primes. Below is an exercise using the factorial concept that will make you doubt that conclusion. Can you figure out what is going on?

   a) What obviously divides

$$5! + 2?$$
$$10! + 2?$$
$$150! + 2?$$

   (Do not actually calculate $n!$ for the above; just write it out in longhand in each instance.)

   b) Without calculating $n!$, indicate a number that obviously divides each number in the following sequence: $7! + 2$, $7! + 3$, $7! + 4$, $7! + 5$, $7! + 6$, and $7! + 7$.

   c) Can you come up with some sequence of numbers that would obviously have *seven consecutive* composites? (Look at (*b*) for a hint.)

   d) Produce a sequence that has ten consecutive composites.

   e) It looks as if we can come up with a sequence that will give us as many composites in a row as we wish. Doesn't this run counter to the claim that there are an infinite number of primes? After a point the primes must stop, if we can get as many composites in a row as we wish. What's wrong with this reasoning?

29. Find a friend and debate the question raised in 28 (*e*) above.

3.2: IN *N:* DISTRIBUTION

The Primes between 1 and n

In subsection 3.1, we carefully developed a procedure to determine the primality of any natural number. Along the way, we explored some interesting byways as well. Suppose we widen our scope a little to ask about primality with regard, not to one number at a time, but to a span of numbers. Suppose we wanted to know, for example, what the primes are between 1 and 72.

At first that would seem to be a foolish question on which to spend much time, since we already have an appropriate procedure to answer it. We can take each number between 1 and 72 and figure out by the procedures discussed earlier whether or not it is prime.

It sometimes happens that by broadening the context within which we ask questions, we gain new insights; this has been an underlying theme in this book as we compared the domains N, T, E, Q, and so forth. But even *within* a domain we can gain new insights if we ask the right questions. When we look at the forest, we sometimes see things that are not apparent with each tree. Furthermore, what is true of the forest may be true not at all of the trees, and vice versa.

There is, in fact, a technique for determining all primes between 1 and n that is simpler than attempting to determine primality for each number individually. The technique is based on one simple observation that is intuitively obvious, together with a strategy we developed in section 3.1.

For the first observation, suppose we know that 3 is a prime. What can we say about multiples of 3 with regard to primality? (See ex. 3.2.1.) They cannot, of course, be prime, since they are all divisible by 3. We can say the same thing about multiples of any other number, even if it is not prime. So if we list all the numbers between 1 and 72 in a row, we could say that though 3 is prime, every third number after that could *not* be prime, thus: 1, 2, ③, 4, 5, 6̸, 7, 8, 9̸, 10, 11, 1̸2̸, 13, ... We circle 3 to remind us that it is prime but cross out every third number. The process can be repeated with 2. We circle 2 and cross out every other number: 1, ②, 3, 4̸, 5, 6̸, 7, 8̸, 9, 1̸0̸, 11, 1̸2̸, ... Notice that some numbers are crossed out in both sequences (what are they?).

Since there is no reason for overkill, we could save some effort and work by first crossing out each multiple of 2, and then starting with 3 (the next number not crossed off) do the same without bothering to cross off what has already been eliminated. We would then have the following: 1, ②, ③, 4̸, 5, 6̸, 7, 8̸, 9̸, 1̸0̸, 11, 1̸2̸, 13, ... What would we do next? Since 5 is the next number not eliminated, it must be prime. (There is an interesting hidden assumption here that we are glossing over. If you are intrigued by what it might be, see ex. 3.2.2.) We circle it and cross off all its multiples.

So far we see the primes between 1 and 72 to be 2, 3, and 5. We have not listed all 72 numbers and so do not yet have a complete picture, but we do know there are many other primes not yet listed (41, for example). But how far do we have to go in crossing off multiples of circled numbers? We know that some of these multiples may have been crossed off earlier, and there is no value in crossing a number off twice or more. Is there some point in the process at which we might be guaranteed that if we stopped crossing off multiples, there would be no danger of having missed actual composites because they would have been crossed off earlier? For example, suppose that in our process we got to the prime number 11 and began crossing off multiples of 11. Those under 72 are 22, 33, 44, 55, 66. Being multiples of 11, they are certainly not prime and do require crossing off, but how do we know whether or not we need 11 to do the initial crossing off? Are there primes smaller than 11 such that their multiples would have eliminated each of these numbers? To see an answer to this question, recall an important observation we made earlier, in section 3.1.

Any composite n must have a prime divisor that is less than or equal to $\sqrt{n}$. So each of the composites listed that are multiples of 11 (up to 66) must have some prime divisor that is less than or equal to the square root of the largest one ($\sqrt{66}$). Since $\sqrt{66} < \sqrt{81} = 9$, there must be a prime divisor for each multiple of 11 in the range that is less than 9. Since 9 is not prime, we need go no higher than 7. This is the second part—f''—of our strategy for finding all primes between 1 and n, as described in section 3.1.

Making Things a Little More Elegant

We can use the process we discussed above to arrive at a more elegant approach to the problem of finding all primes between 1 and n. The procedure was devised by the Greek mathematician Eratosthenes (276–194 B.C.). The refinement he made is essentially in the arrangement of the numbers, which is called the sieve of Eratosthenes. It is depicted in figure 3. What seem to be the advantages (beyond that of having them in a straight line) of arranging the numbers in this fashion?

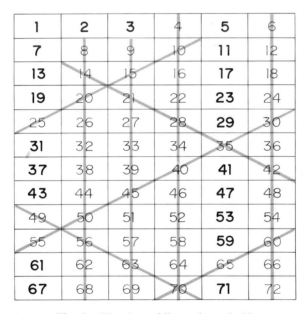

Fig. 3. The sieve of Eratosthenes in N

Notice how easy it is, because of the rectangular format, to eliminate all multiples of any of the primes up to 7. They all lie along convenient lines. The multiples of 2 all lie along three straight lines; those for 3 along two such lines. How would you characterize all the multiples of 7?

The specific rectangular arrangement is not what characterizes the sieve, for we could have arranged things differently, though some arrangements are more convenient than others (see ex. 3.2.4 and 3.2.5).

One interesting observation with regard to the sieve in figure 3 is that with the exception of a few entries in the top row, all primes are in either the first or the fifth column. Can you see why? (See ex. 3.2.6.)

Shifting the Question

Our focus in section 3.2 has been on examining primes between 1 and n. Notice that we have implicitly asked a question in this subsection. That we have done so subtly may be primarily because we have spent so much time answering it rather than trying to clarify what we have asked. Given the title "Primes between 1 and n," what might you ask? What did we ask? Our focus has been on finding primes for any n. We have essentially asked, What are the primes between 1 and n? Thus we have essentially attacked category C on the first page of chapter 3. There are many other questions we could ask (see ex. 3.2.7).

We shall focus on one more question inspired by the following consideration: We know there is no surefire way of *predicting* (as opposed to *testing*) whether a given number in a range will be prime. There is in fact no simple pattern for prime numbers. Despite the fact that we may not be able to find out in an easy way what numbers are prime in an interval between 1 and n (especially if n is enormous), is there any way of indicating *how many* primes are in an interval?

An answer was found independently by two mathematicians in 1896 (Hadamard and Vallée-Poussin), though it had been conjectured by Gauss and Legendre over a century earlier. Furthermore, as with Dirichlet's discovery (see section 2.1), the proofs required machinery from the set of complex numbers, and it was only within the past thirty years that an "elementary" proof was found. The theorem of Hadamard and Vallée-Poussin is now known as the prime-number theorem.

To understand what the theorem claims, it is necessary to know something about *natural logarithms*. By definition, we say that $\log_b a = c$ (read "log of a to base b") provided $b^c = a$. Thus, $\log_{10} 100 = 2$ since $10^2 = 100$. With natural logarithms, the base is e, which approximately equals 2.7 (see ex. 3.2.8 for a delightful practical application that will illuminate what e is about). When the base e is used with logarithms, the logarithm of x is denoted as follows: $\ln x$. Roughly stated, the prime-number theorem claims that provided n is sufficiently large, *the number of primes less than or equal to n—denoted by $\pi(n)$—is approximately equal to* $\frac{n}{\ln n}$.

Let us see what happens, for example, when $n = 6$. What is $\pi(n)$? That is, about how many primes are between 1 and 6? You might be able to find a hand calculator with the ln function to verify that

$$\frac{6}{\ln 6} = \frac{6}{1.7917},$$ which equals about 3.35.

How many primes are between 1 and 6? We can list them; they are 2, 3, and 5. So there are exactly three primes, and since our prediction was about 3.35, we are as close as anyone could hope to be—especially with so small a value for n.

If you have a hand calculator with a natural log function, see what you get for $\pi(n)$ when $n = 100$. That is, find out what $\frac{100}{\ln 100}$ equals. We indicated what $\pi(n)$ was—without using this notation—in table 1, section 2.1, when we first examined the distribution of primes.

Though the proof of the prime-number theorem is considerably beyond the scope of this book, it should not be too difficult—provided you have some understanding of logarithms—to see what the theorem says.

We know that since there are an infinite number of primes, then as n increases without limit, so does $\pi(n)$. Yet we have said that for large n, $\pi(n)$ is very close to $\frac{n}{\ln n}$. In other words, we can get as good an approximation to equality as we wish provided we choose n sufficiently large. How do we build in the notion of near equality of $\pi(n)$ and $\frac{n}{\ln n}$ for large n?

A difficulty in asserting the theorem correctly is that both $\pi(n)$ and $\frac{n}{\ln n}$ become infinitely large, and to say that they both become as large as we wish provided n increases doesn't capture the actual relationship of equality between them.

Before stating the theorem, let us look at an analogy. Suppose we want to compare the relationship between two functions, $2x$ and $4x + 1$ as x increases without limit. It certainly is true that as x increases, both $2x$ and $4x + 1$ increase without limit. Yet it would be deceptive to state that as x increases, $2x$ and $4x + 1$ tend toward equality.

x	$2x$	$4x + 1$	$\frac{4x + 1}{2x}$
1	2	5	2.500
4	8	17	2.125
10	20	41	2.050
100	200	401	2.005
.	.	.	.
.	.	.	.
.	.	.	.

What can we say that is accurate? It is true that as x increases, the *ratio* of $4x + 1$ to $2x$ gets closer and closer to 2.

Many people find it less than natural to compare numbers by examining their ratio, and that is one reason why ideas in the calculus are harder to come by than they might otherwise be. (See ex. 3.2.10 and 3.2.11.)

How would you apply the notion of ratio to the assertion that $\pi(n)$ and $\dfrac{n}{\ln n}$ become close to equal as n becomes infinite? We could say that as n increases, the ratio of $\pi(n)$ to $\dfrac{n}{\ln n}$ tends toward what number? A glance at table 3 will help you see what is involved. Look at the right-hand column. The ratios get closer to what number as n increases?

TABLE 3

As n Increases in the Domain N, $\pi(n)$ Approximates $\dfrac{n}{\ln n}$

n	$\pi(n)$	$\dfrac{n}{\ln n}$	$\dfrac{\pi(n)}{\frac{n}{\ln n}}$
10^1	4	4.343	0.921
10^2	25	21.714	1.151
10^3	168	143.765	1.161
10^6	78 498	72 381.945	1.084
10^9	50 847 478	48 254 630.022	1.053

We can express this fact succinctly by asserting the following:

$$\lim_{n \to \infty} \frac{\pi(n)}{\frac{n}{\ln n}} = 1,$$

where we read the symbol on the left side as "the limit as n goes to infinity of." It is this last statement that is a precise statement of the prime-number theorem.

We saw in the example of the ratio of $4x + 1$ to $2x$ that the ratio approaches 2 as x increases, so that as x increases without limit, it is standard to say that the limit is 2. This means that we can get a ratio as close to 2 as we wish, provided we choose x to be large enough. Can the ratio ever equal 2? As for the ratios in table 3, though they get closer to 1, will they ever equal 1? The question is explored in exercise 3.2.12.

On Integrating Eratosthenes and Hadamard

Though the questions of how to find the primes between 1 and n ((c) in 3.1) and of how many there are between 1 and n ((d) in 3.1) are different, the issues are not independent. Having a procedure for finding the primes (Eratosthenes' sieve) enables us to approximate how many (Hadamard), though the proof of the second from the first is complicated.

A slight modification of the sieve procedure (how to find) yields another "how many" analysis that is a little surprising. Recall that in Eratosthenes' procedure, we first cross out all multiples of 2, then all multiples of the next number—3—that were not eliminated, then all multiples of the next number—5—that were not eliminated, and so forth. When we try to cross off all multiples of 3, we find that half of them (see fig. 3) have already been eliminated (why?), and so we end up crossing out every sixth number rather than every third number.

Hawkins (1958) conducted an interesting variation of this procedure that you might be able to simulate on a computer or perhaps by some other random procedure (such as drawing numbers out of a hat). Given an array (for example, the one in fig. 3), he first circled the number 2, and then instead of eliminating every *other* number from the list, he used a procedure for randomly eliminating *half* the remaining numbers (but not necessarily every other one). He then took the next number after 2 that had not been eliminated. Suppose it was the number 4. He circled 4, treating it as the next prime beyond 2, since it had not been "sieved out." He then eliminated not every *fourth* number, but *one-fourth* of the remaining numbers (allowing for duplication from those that were already eliminated) *at random*. He then chose the next number that had not been eliminated. Suppose it was 8. What portion of the remaining numbers would he cross off at random? (See ex. 3.2.13 for a discussion of how to do this.) He stopped the procedure when he got to a number not sieved that was about equal to $\sqrt{n}$.

He took all the circled numbers—which he called "random primes"—and what do you think he found? He was able to show that the number of these "random primes" was approximately equal to $\dfrac{n}{\ln n}$—just as with actual primes! (See ex. 3.2.14.)

On Clustering and Doodling

In section 3.2 we discussed procedures for finding primes in an interval as well as a method for finding out how many there are. In addition, we have shown, through the "random prime" procedure, one surprising way in which the two ideas connect. Though the sieve in figure 3 shows that all primes (except some in the front row) appear in only two of the columns, that result is a consequence of the way we chose to arrange the elements of N (that is, in six columns).

Despite the presence of almost all primes in the first and fifth columns, nonprimes obviously appear there as well. That is not surprising, for as we have discussed throughout, there is no very nice and predictable pattern in which primes display themselves in N. Given this lack of predictability, it is particularly illuminating for us to be able to exhibit clusterings of primes in any interval. We are thus quite happy to glorify tendencies even if they are only partial.

A fascinating clustering of primes has recently been discovered by S. M. Ulam and his associates at the University of California (Stein, Ulam, and Wells 1964). What is particularly interesting about their finding is that it resulted from mere "doodling." How many of you have spent time drawing sketches like these:

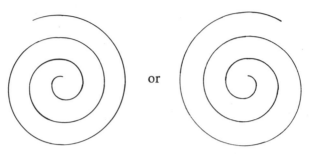

Ulam began his research in essentially that way. He decided, however, to superimpose the elements of N in a systematic way on one of the spirals above, which led to some results that were truly eye-opening. He essentially did the following:

$$\begin{array}{cccc} & & 13 & \\ 5 & 4 & 3 & 12 \\ 6 & 1 & 2 & 11 \\ 7 & 8 & 9 & 10 \end{array}$$

The picture will be easier to analyze if we straighten things out a little:

$$\begin{array}{|cccc|} \hline & & & 13 \\ 5 & 4 & 3 & 12 \\ 6 & 1 & 2 & 11 \\ 7 & 8 & 9 & 10 \\ \hline \end{array}$$

To make sure that the numbers are equally spaced, we shall put the affair on graph paper as shown here:

$$\begin{array}{|cccc|} \hline & & & 13 \\ 5 & 4 & 3 & 12 \\ 6 & 1 & 2 & 11 \\ 7 & 8 & 9 & 10 \\ \hline \end{array}$$

What might you do with such a picture? Well, pictures in themselves may be beguiling, but they do not ask questions. There are many valuable and interesting questions you can ask about the array above. (Before proceeding, try ex. 3.2.15 to limber yourself up.)

One thing Ulam did that encouraged a search for patterns that eventually turned out to yield a considerable payoff was, not to focus on the spiral effect, but to draw diagonals (at 45° angles to the horizontal-vertical effect of the graph) and to examine the numbers that lie along these diagonals.

A schematic rendering of his musings appeared on the cover of *Scientific American* in 1964 and is reproduced in figure 4 (see Gardner [1964]). Look along some of the diagonal lines sketched in and what do you notice? Perhaps many things stand out, but if you keep in mind the concept of prime numbers, what do you observe?

100	99	98	**97**	96	95	94	93	92	91
65	64	63	62	**61**	60	**59**	58	57	90
66	**37**	36	35	34	33	32	**31**	56	**89**
67	38	**17**	16	15	14	**13**	30	55	88
68	39	18	**5**	4	**3**	12	**29**	54	87
69	40	**19**	6	**1**	**2**	**11**	28	**53**	86
70	**41**	20	**7**	8	9	10	27	52	85
71	42	21	22	**23**	24	25	26	51	84
72	**43**	44	45	46	**47**	48	49	50	**83**
73	74	75	76	77	78	**79**	80	81	82

Fig. 4. A spiral grid indicating the distribution of primes in N (from cover [Prime-Number Pattern] March 1964 issue; copyright © 1964 by Scientific American, Inc.; all rights reserved)

1. Some diagonals have *no* primes at all—for example, 10, 28, 54, 88.
2. Some diagonals have few primes—for example, 11, 3, 15, 35, 63, 99.
3. Some diagonals are very rich in primes—for example, 7, 1, 3, 13, 31, 57, 91.

We have, of course, displayed only a small part of each infinite diagonal and therefore need to be tentative in our generalizations—well, almost tentative. Can you make any statement of which you are pretty certain with regard to the three categories mentioned above and a particular diagonal? (See ex. 3.2.16.)

In order to extend their field of vision, Ulam and his associates placed the points on an electronic computer with a visual display. The locations of primes were programmed to light up. Figure 5 represents the visual display. Just a glance at that display shows rich prime lines that seem to pop out. The authors comment, "It is a property of the visual brain which allows one to discover such lines at once and also notice many other peculiarities of distribution of points in two dimensions. In a visualization of a one-dimensional sequence this is not so much the case (perhaps an acoustic interpretation would be more suggestive)" (Stein, Ulam, and Wells 1964, p. 517).

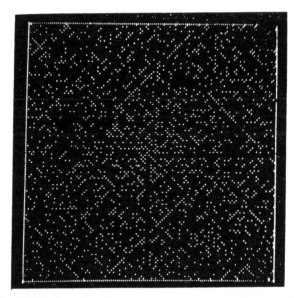

Fig. 5. The distribution of primes on a computer (reprinted, with permission, from *American Mathematical Monthly* 71 [May 1964] :517)

Once Ulam saw the visual display (which was capable of exhibiting up to 65 000 points), he searched for a way to characterize those "prime rich" lines in the form of an equation and then investigated up to ten million primes with a computer. He found that of some diagonal lines almost *half* the entries were prime, whereas others had almost none. Let us hint at how he went about exploring prime-rich and prime-poor numbers along diagonals. Look at the diagonal 1, 3, 13, 31, 57, 91. What do you notice? Again, many things may pop out, for example, that the numbers are all odd, or that none of the numbers end in 5. One observation is particularly relevant here. Notice the *differences* between terms:

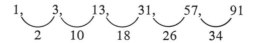

What do you see? Let's put them on a line by themselves: 2, 10, 18, 26, 34.

We have encountered numbers like these before. They form an arithmetic progression (the difference between terms being 8). What's so great about the differences between terms forming an arithmetic progression? Before answering that question, explore other diagonals and see whether or not you get the same pattern (see ex. 3.2.17).

A Helpful Digression

There is a fascinating property of equations that you may not have come across before, though you have already met the converse of the notion in a problem set. We shall develop the idea and then apply it to the spiral.

Take any linear equation, like $y = 2x + 5$. Suppose you let $x = 1, 2, 3, 4, 5$. We notice that the y values are 7, 9, 11, 13, and 15 respectively. The y values form an arithmetic progression if the equation is a linear one. The differences between the y values are constant (2 in this case). Persuade yourself, first of all, that this is not a fluke by choosing another linear equation of your own and substituting 1, 2, 3, and 4 for x. (See ex. 3.2.18 for a discussion of why, in general, this is so.)

Suppose you take a quadratic equation. Substitute $x = 1, 2, 3, 4$, and so on, and see what happens. Actually, we have done so for a special case in exercise 2.1.8. Let us review it here:

$$y = x^2 + x + 41$$

Look at part (*b*) of the exercise. We found the differences between corresponding y values to be 4, 6, 8, and 10. These form an arithmetic progression. Was this a fluke too? Take your own quadratic equation and substitute $x = 1, 2, 3, 4, \ldots$, and see what happens to the y values. (See ex. 3.2.19 for a discussion of why this is generally so.)

To say that the y differences form an arithmetic progression implies that in a quadratic equation the second differences are constant. We already noticed in a linear equation that the *first* differences (the values of y directly) are constant. It is a general property of equations, which you might wish to explore, that if an equation is of the nth degree, the nth differences are constant (try it for a third-degree equation). The point we will use, without proof, is the converse of this phenomenon. That is, if we have a sequence, and the nth differences of the sequence are constant, then we can find a simple equation of the nth degree that would pass through the original points.

Our interest is in the diagonals of the spiral in figure 4 (which you may easily have forgotten after this long digression). It turns out that the elements along the diagonal we are focusing on form an arithmetic progression. The second differences are therefore constant, and we thus should be able to find a quadratic equation to define these points. We shall now return to the spiral.

Back to the Spiral

Look again at 1, 3, 13, 31, 57, 91. Since the second differences are constant, there must be a quadratic that defines it. What is that quadratic? We know that the general form of a quadratic equation is $y = a \cdot x^2 + b \cdot x + c$. Moreover, we know that—

- for the first value of x, $y = 1$;
- for the second value of x, $y = 3$;
- for the third value of x, $y = 13$;
- for the fourth value of x, $y = 31$;
  and so forth.

How do we find the values for a, b, and c that work for this quadratic? We merely plug in the values of x and y that correspond to each other and pray! Let's see:

If $x = 1$, then $y = 1$, so the quadratic equation takes this form:
$$1 = a \cdot 1^2 + b \cdot 1 + c$$
If $x = 2$, then $y = 3$, so the quadratic equation takes the form:
$$3 = a \cdot 2^2 + b \cdot 2 + c$$
If $x = 3$, then $y = 13$, so
$$13 = a \cdot 3^2 + b \cdot 3 + c.$$

We could add more equations, but let's summarize and systematize:
$$1 = a + b + c$$
$$3 = 4a + 2b + c$$
$$13 = 9a + 3b + c$$

Here we have three equations in three unknowns, and we can solve them by focusing on two pairs (see ex. 3.2.20 for the calculations). We end up with $a = 4$, $b = -10$, $c = 7$, and thus the desired quadratic is
$$4 \cdot x^2 - 10x + 7.$$

Take a few other diagonals with constant second differences and see if you can find quadratics for them.

What is the advantage of thus characterizing a diagonal that looks rich in primes? With appropriate machinery (as Ulam had), we can gain some insight into whether rich-looking diagonals really have a high concentration of primes or whether the few that we do observe give a misleading impression. Just as important, we can take poor-prime diagonals and see if they are prime poor only in certain segments. It should be possible for you to prove that some diagonals can be entirely devoid of primes by using quadratic equations to characterize the diagonals (see ex. 3.2.21).

In Defense of Doodling

It turned out that Ulam's doodling yielded a significant payoff by enlarging our view of prime-rich lines. Even if such doodling had not led to connections with primes, however, we still would have had an enjoyable excursion into quadratic forms (as well as other ones), prompted by the search for a formula to characterize numbers along the diagonals. Of course, doodling in itself would have accomplished very little. We must both doodle and attend to our doodling in order to find "gold" where we saw only "dross" before. But how do we analyze doodling?

Before pursuing the question, let us look at doodling in a larger context. Doodling represents a playful spirit in which we allow free rein to our imagination as well as our pencil. We allow ourselves, for a while, to think things that are "senseless" and to do things that may be "worthless."

How do we make sense of our doodling? We become conscious of what we are doing and learn to impose form on our doodle. (What form did Ulam impose?) In addition to imposing form on what we are doing, we must ask ourselves questions. After asking questions (though sometimes prior to it), we could *modify* the doodle.

Suppose you wanted to extend Ulam's doodle. What might you do?

1. You could focus on lines other than the diagonals of 45° inclination. What lines might you choose?
2. You could select lines other than straight lines to examine.
3. You might form a new rectangular array using some scheme other than a spiral, but still keep a rectangle-shaped scheme. Does the following suggest a possibility?

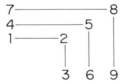

Having formed such an array, you could choose diagonals, vertical lines, horizontal lines, and others to focus on. But what does "focus on" mean? Among other things it means asking questions. What questions might you ask in each of the instances above?

(In the Bibliography are listed several pieces written by this author together with Marion Walter that generalize the notion of doodling and that suggest strategies for finding and seeing enchantment and wonder in almost all mathematical thought.)

What Ulam produced is beautiful, but each of us is capable of generating such wonder out of essentially nothing—provided we have the confidence and the courage to ask offbeat questions about the most mundane activities.

PROBLEM SET 3.2

1. How would you demonstrate in the manner of figures 1 and 2 of section 1.2 that multiples of primes cannot be prime?

2. We have claimed that since 5 was *not* crossed out by any of the previous primes, it must be a prime. There are several assumptions actually being made here, the first, raised earlier in this chapter, is "silly." Partly to encourage you to raise silly questions of your own, we raise it again:

   a) How do you know that it is not divisible by some prime that comes later? What assumptions about natural numbers would be violated if that happened? This is a nice question for debate.

The second question is less silly; it also came up before in an exercise in section 2.1. If the question interests you, you might wish to look back at the exercises in section 2.1.

   b) Suppose we know that since 5 has not been crossed out, there is no prime that divides it. How do we know that there is no composite that divides it? In this case, 4 is the only composite that precedes it, and it is easy to verify that it does not divide 5. But the situation could have been different. No?

3. In the process of crossing out composites between 1 and 72, we looked at multiples of 11 and claimed that they would all have been crossed out already. Show what primes before 11 would have enabled us to cross out 22, 33, 44, 55, 66. Things couldn't be too complicated because the only choices are 2, 3, 5, 7.

4. Arrange a sieve for determining all primes between 1 and 72 using a table that has 5 columns (notice that you will not have all boxes filled). Compare that sieve with figure 3 for ease of crossing out composites.

5. Make a sieve for numbers between 1 and 200. Have a friend choose one rectangular array, and choose another yourself. Which array do you find easier? Why? Discuss this question with a friend.

6. In figure 3, with the exception of the first row, all primes are in either the first or fifth column. To see why this is so, look at the entries in the last column. They all can be characterized as numbers of the form $6n$. Using this characterization as a reference point, how would you characterize all entries in the first column? The third column? You have seen this concept before in section 2.1, and you might want to reexamine arithmetic progressions and Dirichlet's find. How many primes will there be in the first column? Is it possible that primes would eventually appear only in the fifth column?

7. Given a concept or category, we frequently impose unnecessary limits on the kinds of questions we might ask about it. For example, given

the category of primes between 1 and n, we investigated first, "What are those primes?" and "How do you find them?" We might ask many other questions besides the one in the text that provided our second point of focus, "How many are there?" Find a friend or two, and try, without attempting to analyze answers, to come up with as many questions pertaining to the category of primes between 1 and n as you can.

8. The following delightful sequence represents a disguised introduction to a definition of e, the base of natural logarithms.
   a) Suppose you put $1.00 in the bank at 100% interest compounded annually. How much money will you have at the end of the year?
   b) If the interest is compounded, how much will you have at the end of two years, provided the rate is 50% a year?
   c) How much at the end of four years if the interest rate is 25% a year?
   d) In general, if you place P_1 dollars in the bank and you compound it annually at r% interest, then at the end of one year you will have $A_1 = P_1 + \frac{r}{100} P_1$. At the end of two years you will have—

$$A_2 = A_1 + \frac{r}{100} A_1 = A_1\left(1 + \frac{r}{100}\right) = \left(P_1 + \frac{r}{100} P_1\right) \cdot \left(1 + \frac{r}{100}\right) = P_1\left(1 + \frac{r}{100}\right) \cdot \left(1 + \frac{r}{100}\right) = P_1\left(1 + \frac{r}{100}\right)^2.$$

At the end of three years, you will have—

$$A_3 = P_1\left(1 + \frac{r}{100}\right)^3.$$

In general, at the end of n years, you will have—

(β) $$A_n = P_1\left(1 + \frac{r}{100}\right)^n.$$

Suppose that you put a dollar in the bank, and the bank has the following unlikely table for determining interest rate:

Number of Years	Compound Annual Interest Rate
1	100%
2	50%
3	33⅓%
4	25%
.	.
.	.
.	.
n	$\left(\frac{100}{n}\right)$%

In questions (*a*) through (*c*) you already figured out answers to some interest questions that are related to this table. Under such a scheme how much would you have in the bank after ten years if you began with a dollar? You might have the impression that if you left the money in long enough (and if you lived long enough) you could become a millionaire. This is one of many problems in which our intuition deceives us.

The general formula (using β) for the amount of money due us after n years is the following: $A_n = 1\left(1 + \frac{1}{n}\right)^n = \left(1 + \frac{1}{n}\right)^n$. If you have worked with logarithms before (otherwise the amount of calculation required is burdensome), you can figure out that after 100 years you will have $\left(1 + \frac{1}{100}\right)^{100}$, which equals approximately \$2.70.

If you know how to use logarithms, figure out how much you would have after 1000 years? Are you surprised? It turns out that no matter how large n is, $\left(1 + \frac{1}{n}\right)^n$ is always less than 3. As a matter of fact, one definition of e is, Take the limit of $\left(1 + \frac{1}{n}\right)^n$ as n gets as large as you wish (to eight decimal places, $e = 2.7182818$).

9. There is another definition for e (and in a more advanced course it would be proved equivalent) that may interest you. It is

$$e = 1 + \frac{1}{1!} + \frac{1}{2!} + \frac{1}{3!} + \frac{1}{4!} + \cdots$$

Can you persuade yourself that this expression is less than 4?

10. In comparing the functions $4x + 1$ and $2x$ as x increases, we found it helpful to look at their ratios; they approach 2. How might you compare these two functions other than by ratio? One obvious way is to compare their differences. See what happens when you focus on the differences as x increases. What kinds of statements can you make?

11. Given any two numbers, we could compare them by examining their ratio or by examining their difference. Come up with several examples to show when the concept of ratio makes sense in comparing numbers and when the concept of differences makes sense. Are there other ways you might compare numbers? This would be a valuable topic to discuss with a friend.

12. We inquired whether any entry in the last column of table 3 will ever equal 1. It turns out that $\ln n$ is always irrational (cannot be expressed as a ratio of elements of N) for $n \in N$. Though we shall not prove that $\ln n$ is irrational for n belonging to N, we can demonstrate that if we accept this,

then our claim follows easily. We will show that the irrationality of $\ln n$ implies the irrationality of $\dfrac{n}{\ln n}$, which in turn implies the irrationality of
$$\dfrac{\pi(n)}{\dfrac{n}{\ln n}}.$$

First let us show that irrationality of $\ln n$ implies irrationality of $\dfrac{n}{\ln n}$. Let $\ln n = c$. If c is irrational, then we cannot express it in the form $\dfrac{a}{b}$ for a, $b \in N$. Suppose that $\dfrac{n}{\ln n}$ were rational (again we are using a proof by contradiction). Then $\dfrac{n}{c} = \dfrac{e}{f}$ for some $e, f \in N$. But then $c = \dfrac{n}{\frac{e}{f}} = n \cdot \dfrac{f}{e} = \dfrac{n \cdot f}{e}$. Why? But since $n, f, e \in N$, $\dfrac{n \cdot f}{e}$ is rational, and we contradict the assumption that c is irrational. Now prove that $\dfrac{\pi(n)}{\frac{n}{\ln n}}$ is irrational. (*Hint:* you need not work hard.)

13. Make a chart like figure 3 using the natural numbers 1 to 54. Using a sieve procedure like that of David Hawkins, place the integers 3 through 54 in a hat and mix them up. Draw out half (26 in this case), and cross these off the sieve. Find on the chart the next number after 2 that has not been crossed off the list; call it m. Remove this number and all numbers that come before it, so they cannot be drawn from the hat. Put all other integers *back* into the hat. Now divide the number of integers that are back in the hat by m and apply the greatest-integer function to your quotient. Remove that many integers from the hat. (*Example*: If the next number after 2 not crossed off the list is 5, you would make sure that 3, 4, and 5 are not put back in the hat. Now there are forty-nine integers in the hat. Since $\dfrac{49}{5} = 9\dfrac{4}{5}$, we draw out 9 integers.) Cross these integers off the sieve, *allowing for duplication* of those already eliminated. Repeat the procedure, that is, find the next number m that has not been crossed off, remove it and all integers preceding it; put the others back in, draw $\left[\dfrac{54 - m}{m}\right]$ numbers from the hat, and cross these off. Stop the procedure when you get to the largest m that is not greater than 54 (in this case, the largest sieve number you could have is 7, since $7 < \sqrt{54} < 8$). How many numbers have not been sieved? How many primes are there in figure 3 (up to 54)? Compare the results. Try the experiment about ten times and average your results. How close do you come to the actual number of primes? (I am grateful to

73

Margaret Stempien for rewriting the question to clarify the meaning. She was a student in my course at State University of New York at Buffalo in 1978.)

14. According to the prime-number theorem, how many primes would you expect in figure 3? How many do you get? Are the numbers of "random primes" you get closer to the actual number of primes in figure 3 or to the number predicted by the formula?

15. Given Ulam's graph-paper display of the numbers 1–13, what kinds of questions might you ask? If you simply follow the spiral, you will see little more than the numbers 1 through 13 in order; try, therefore, to "break through" the spiral by relating numbers as they appear in the spiral in nonsequential ways.

16. You are pretty well assured that a diagonal in figure 4 will have no primes if all the elements on the diagonal are even numbers (excluding the number 2). Are there any such diagonals? Can you convince yourself that there will never be an odd element on those diagonals that you have singled out? How? This would be a nice exercise to do with someone else. You will probably each interpret and justify what is going on differently.

17. In searching for a pattern of arithmetic differences between numbers along a diagonal in figure 4, notice that with regard to a horizontal line like 1, 2, 11, 28, 53, 86, we can—

   a) focus on the numbers along a diagonal *above* that horizontal;

   b) focus on the numbers along a diagonal *below* the horizontal;

   c) attempt to jockey back and forth.

For example, in the text we explored 1, 3, 13, 31, ... (option *a*). We could have explored either 1, 7, 21, 43, ... (*b*) or 1, 3, 7, 13, 21, 31, ... (*c*). Can you get an arithmetic progression for the jockeying (*c*) of 1, 3, 7, 13, 21, 31, ... ? Try others and see what happens with (*b*) or (*c*) strategies.

18. Why is there a constant difference between the *y* values of a linear equation? You can look at the problem both algebraically and geometrically.

   a) Algebraically

   $y = mx + b$ is the equation of a straight line. Substituting elements of N for x we get the following:

   $$y_1 = m(1) + b = m + b$$
   $$y_2 = m(2) + b = 2m + b$$
   $$y_3 = m(3) + b = 3m + b$$
   $$y_4 = m(4) + b = 4m + b$$

   It is clear that $y_n - y_{n-1} = m$ for all cases.

b) Geometrically

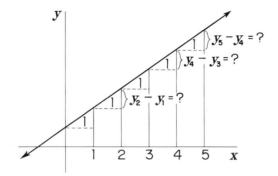

Relate your thinking here to a key characteristic of straight lines. What is being claimed about the y differences? Why is it true?

19. Why do we get an arithmetic progression for y differences when we plug in $x = 1, 2, 3, 4, \ldots$ for a quadratic equation? The general form of a quadratic is $y = ax^2 + bx + c$; plug in $x = 1, 2, 3, 4$, and see what happens. Those of you who have studied calculus might find another explanation.

20. How do you solve the following equations:

a) $1 = a + b + c$
b) $3 = 4a + 2b + c$
c) $13 = 9a + 3b + c$

One way is as follows: If we subtract (a) from (b) we get

$$3 = 4a + 2b + c$$
$$1 = a + b + c$$
$$2 = 3a + b$$

d)

If we subtract (b) from (c) we get

e) $10 = 5a + b$.

Now focus on (d) and (e). Subtract (d) from (e):

$$8 = 2a$$
$$a = 4$$

Once we know $a = 4$, it is easy to see that $b = -10$ and $c = 7$. Verify that the quadratic $y = 4x^2 - 10x + 7$ is correct by substituting the fourth value of x along the diagonal and seeing what you get for y.

21. Prove that the diagonal 2, 12, 30, 56, 90 has no primes other than 2 by finding a quadratic for these numbers. Before coming up with the quad-

ratic, what can you guess about the relationship of *a*, *b*, and *c* that would account for the poor performance of primes?

3.3: IN *E*: DETERMINATION AND DISTRIBUTION

Determination

We have covered considerable ground in examining questions regarding the determination and distributions of primes in *N*. The amount of thought that must have been invested over the years in these questions is truly mind boggling. Let us now consider similar questions with reference to the domain *E*.

We have suggested in section 2.2 that if a number is of the form $2 \times \sigma$ (for σ any odd element in *N*), then it must be a prime. Our object was to come up with a generating formula. Here was shall use the same observation to uncover a simple test for primality. It is worth recalling that though in *E* the association of $2 \cdot \sigma$ with primality was not airtight, a great deal of experience with primes in *E* suggests that the association is a valid one.

For a number to qualify for a test of primality, we need to show not only that if it is of the form $2 \cdot \sigma$, then it must be prime in *E* (for there may be other primes in *E* that are of different forms), but also that if it is *not* of that form, then it is *not* prime. (If you are puzzled by why both conditions are required, see ex. 3.1.22, 23, 24, in which we discuss what is required to test for divisibility by some numbers.)

We discuss in exercises 3.3.1 and 3.3.2 how it is that both these conditions are satisfied, and you may either accept this test for primes in *E* or refer to those arguments. You will most likely find these proofs less interesting than what follows, and since you need not be more than reasonably sure that the test works to follow the arguments in this section, your order of priority should depend on whether you prefer dessert before or after your main course.

Thus, one possibility is to check to see if the alleged prime in *E* is of the form $2 \cdot $ odd. For example, is 72 a prime in *E*? To see, let us divide by 2. We do so and get 36. Since 36 is even, 72 is not prime, Try 102; $\frac{102}{2} = 51$, and since 51 is odd, the number 102 is prime.

That's one way of testing. Let's look for another. We showed in section 2.2 that the condition of $2 \cdot \sigma$ can be easily transformed to $4n - 2$. That is, an element of *E* is prime if and only if it is of the form $4n - 2$. Well, how can we tell if it is of that form? Look carefully at the form $4n - 2$. All elements of *N* are expressible in one of the following forms: $4n$, $4n - 1$, $4n - 2$, $4n - 3$ for $n \in N$ (you may wish to look back at section 2.1, on arithmetic progression).

What about the elements of *E*? Two of these four forms must be eliminated (why?). The only possible forms in E are $4n$ and $4n - 2$ (discounting what special case?). Now, given any element of E, how can you tell whether it is of the form $4n$ or $4n - 2$? Since 4 obviously divides the first form, and 4 cannot divide the second, we have a very simple test (see ex. 3.1.3 if you are in doubt).

If 4 does not divide the alleged prime in E (discounting what element again?), then that number is prime. Is 72 prime? $4|72$, so it is not. Is 102 prime? $4\!\not|\,102$, so it is prime! It is necessary to bear in mind that the division takes place in N.

You might now enjoy returning to some of the generating-formula questions in exercises from section 2.2. For example, how do we know that the modified generating formulas of Mersenne, Fermat, and Euler, which we came up with in section 2.2, always yield primes in E? (See ex. 3.3.5 and 3.3.6, though it may be better to wait until you have read the discussion below before proceeding.)

Let us look back at Dirichlet's generators that we modified for E. Recall that in N, if a and d are relatively prime, then $a + dn$ will generate an infinite number of primes (though many composites as well).

In examining the analogous problem in E, we found that a and d need not be relatively prime for $a + dn$ ($n \in E'$ this time) to yield an infinitude of primes. For example, $2 + 8n$ seems to generate an infinitude of primes (18, 34, 50, 66, ...), and 2 and 8 are not relatively prime. On the other hand, if they *are* relatively prime, then we do not necessarily get an infinitude of primes. We can see that 8 and 6 are relatively prime in E, and yet $8 + 6n$ seems to generate no primes. We may sometimes, however, get an infinitude of primes when a and d are relatively prime. For example, consider the progression for $6 + 8n$. With our divisibility test for primes in E, it is not difficult to see what is going on.

Look at $8 + 6n$. Let us reduce testing to the concept of divisibility in N. Factored, $6 = 2 \cdot 3$, and since $n \in E'$, $n = 2 \cdot m$ for some $m \in N$. Therefore, $6 \cdot n = 2 \cdot 3 \cdot (2 \cdot m) = 4 \cdot (3m) = 4 \cdot k$ for some k. Then $8 + 6n = 8 + 4k$. Now, resorting to our handy divisibility test, we find that 4 divides $8 + 4k$ no matter what k is in N, and we are done, for any number that 4 divides cannot be prime.

Why is the issue so different with $6 + 8n$? It is obvious that $4|8n$ regardless of n, and we also see that $4\!\not|\,6$ (4 does not divide 6). Why, then, can we conclude that $4\!\not|\,(6 + 8n)$ for any $n \in E'$? If we can see that $4\!\not|\,(6 + 8n)$ for any $n \in E'$, it becomes clear that all elements of $6 + 8n$ are prime.

Suppose we look separately at a and dn in a modified version of Dirichlet's formula in E. Since both d and n come from E', what can you say about dn with regard to divisibility by 4? That $4|dn$ always. Does $4|a$ always if $a \in E'$? Obviously not. We are now ready to make a conjecture regarding the generation of primes in E through the use of a modified version of Dirichlet's formula.

If $a + dn$ is such that a is prime in E, then Dirichlet's formula will generate an infinitude of primes and no composites. On the other hand, if a is composite, then the formula generates *no* primes. We meet here a phenomenon that we shall see much more of in the second portion of 3.3. We are reminded of the famous show tune "With Me It's All or Nothing."

Distribution

In this section, we shall focus on three interesting distribution questions in E that came up in N: those raised by Eratosthenes, Hadamard, and Ulam. As you are perhaps beginning to expect, the analogous results in E will have greater regularity and thus less interest than those of N. (See exercise 3.3.9 for a fascinating question regarding this assertion.) We shall, however, find some surprising results now and again (both here and in chapters 4 and 5), so don't despair. Besides, we shall use some interesting mathematical ideas to verify the regularity.

Eratosthenes Revisited

We turn first to the sieve of Eratosthenes. Let us create one that is analogous to the one we used in N. If we have six columns and twelve rows, we fill it in with the numbers from 1 to 142 (fig. 6).

1	2	4	6	8	10
12	14	16	18	20	22
24	26	28	30	32	34
36	38	40	42	44	46
48	50	52	54	56	58
60	62	64	66	68	70
72	74	76	78	80	82
84	86	88	90	92	94
96	98	100	102	104	106
108	110	112	114	116	118
120	122	124	126	128	130
132	134	136	138	140	142

Fig. 6. The sieve of Eratosthenes in E

As with the sieve in N, we choose the first prime, 2, and then cross off all multiples of 2. But what are the multiples of 2 in E? The first one is 2×2, the second 2×4, the third 2×6, and so forth. Why is 2×3 not a multiple of 2 in E? If 2×3 were a multiple, then 2 would have to divide 2×3—which it cannot do in E.

Let us now move on to the next number that has not been crossed out, 6, the next prime in E. Begin to cross out all multiples of 6: 6×2, 6×4, $6 \times 6, \ldots$ but wait! They've been crossed out already. Can you see why? Notice that with a single stroke all the composites have been sieved out in E.

In addition, it is interesting to note that all the composites are in the first, third, and fifth columns and all the primes in the others. You might wonder why this has occurred. It turns out that (with the exception of the number 1), any entry in a column differs from any other entry below it in that column by a multiple (in this case, meaning "multiple in N") of 12. If we focus on the first element in the top of the column, then we can express any other element in the form $a + 12n$ for $n \in N$. For example, 42 is expressible as $6 + 12 \cdot 3$, where 6 is the top number in the column. Furthermore, 52 is expressible as $4 + 12 \cdot 4$. Notice that as goes the number on top of each column (with the exception of 1), so go all the numbers in the column. Do you see why? Does an expression of form $a + 12n$ look familiar? (See ex. 3.3.12.)

We are back to the observation we made in the previous section with regard to Dirichlet. We now turn to our second distribution question.

Hadamard Revisited

What becomes of the prime-number theorem in E? That is, can we find a way of approximating the number of primes between 1 and e for e any element of E? Let us look again at some of the elements of E and underline the primes:

$$\{1, \underline{2}, 4, \underline{6}, 8, \underline{10}, 12, \underline{14}, 16, \underline{18}, \ldots\}$$

Starting with 2, every other element in E is prime. This means that, as a good first approximation for any element e of E, about half the numbers between 1 and e in the set are prime. Therefore, if we wanted to find the number of primes between 1 and e, then we might choose $1/2 \cdot e$ as a good approximation. Suppose e were 14. Then we would claim that there are about seven primes in E between 1 and 14. Actually, there are only four. What happened? Although about every other number is prime in E, half the counting numbers (the odd ones) do *not* appear in E at all. Thus, $1/2 \cdot e$ would give too high a number, for we have to eliminate half the elements of N that never appear in E.

We thus would choose as our approximation $1/2 \, (1/2 \cdot e) = 1/4 \, e$. This should tell us approximately how many primes there are between 1 and e for any element in E. Let's see how accurate we are. The expression $\pi(e)$ (the same symbol used in N) stands for the number of primes between 1 and e in E:

$$\pi(4) = \frac{1}{4} \cdot 4 = 1$$

$$\pi(8) = \frac{1}{4} \cdot 8 = 2$$

For reasons that are easy to see, we couldn't be happier! If e is a composite (multiple of 4 in N), then $1/4 \cdot e$ yields the number of primes exactly!

What a windfall. We saw that in N, the best we could get was an approximation for $\pi(n)$. Here we get a number that will exactly equal $\pi(n)$ in every instance rather than approximate $\pi(n)$ for large elements of E only!

Before we get carried away with ourselves, however, we ought to do several things. First of all, let's try a few more composites to persuade ourselves that our discovery indeed always holds. Second, let's explore what happens with primes in E. Let's try our formula on 14 and 18:

$$\pi(14) \stackrel{?}{=} \frac{1}{4} \cdot 14 = 3\frac{1}{2}$$

$$\pi(18) \stackrel{?}{=} \frac{1}{4} \cdot 18 = 4\frac{1}{2}$$

But we know that $\pi(14) = 4$ and that $\pi(18) = 5$.

It appears, then, that we are off by $1/2$, if the number in E is prime. Try a few more examples on your own to persuade yourself that if p is a prime in E, then $\pi(p) = 1/4 \cdot p + 1/2 = \dfrac{p+2}{4}$.

We thus have two cases to consider, summarized below:

$$\pi(e) = \begin{cases} \dfrac{1}{4} \cdot e & \text{if } e \text{ is composite} \\ \dfrac{e+2}{4} & \text{if } e \text{ is prime} \end{cases}$$

We are almost done, though we have obviously forgotten the most ornery element of E. What if $e = 1$? We might try the two formulations we have so far on 1:

$$\pi(1) \stackrel{?}{=} \frac{1}{4} \cdot 1 = \frac{1}{4};$$

$$\pi(1) \stackrel{?}{=} \frac{1+2}{4} = \frac{3}{4}.$$

It is obvious that neither of the equations above works, for we know that $\pi(1) = 0$. We could accurately summarize our results as follows:

$$\pi(e) = \begin{cases} 0 & \text{if } e = 1 \\ \dfrac{1}{4} \cdot e & \text{if } e \text{ is composite} \\ \dfrac{e+2}{4} & \text{if } e \text{ is prime} \end{cases}$$

We have, then, three separate rules to follow, depending on the nature of e in E. Considering the fact that if done properly we get an exact answer every time, that's a small price to pay. Still, it would be nice if there were some succinct way of finding $\pi(e)$ for e any element in E. Since you will have a chance in the problem set to analyze the situation on your own for an analogous domain, your pleasure will be spoiled only slightly by a continued analysis of this problem.

What we want is a function that will (1) allow us to add $1/2$ to $1/4 \cdot e$ in order to get a next whole number when e is prime but (2) ignore a $1/2$ that we add on when e is composite and (3) permit the reduction of $1/2$ added to $1/4 \cdot 1$ to be 0 when $e = 1$.

We encountered a function in 2.1 that would be appropriate. It was involved in Mills's humorous answer to the question of finding a generating formula for primes in N. If we focus on e when e is prime, then what function of $\dfrac{e+2}{4}$ will accomplish our objective above? The square bracket function ($[x]$ = greatest integer less than or equal to x) will do the trick, for

$$\left[\frac{e+2}{4}\right] = \begin{cases} 0 \text{ if } e = 1 \\ \dfrac{e+2}{4} \text{ if } e \text{ is prime} \\ \dfrac{e}{4} \text{ if } e \text{ is composite} \end{cases}.$$

Ulam Revisited

What becomes of Ulam's magnificent doodling in E? Let us take figure 4 and impose the elements of E on it in the same spiral fashion we did for N. The result is figure 7. What do you notice? We see a striking similarity with the spiral in N in the configuration of its obvious failures. That is, as with N, there are failures along every other diagonal. But the successes are more spectacular. Along any diagonal where there are successes, there are *no* failures. Again, we're back to the theme song, "With Me It's All or Nothing."

Why is there such a sharp distinction between poverty and wealth here? There are many ways of seeing why this pattern will continue, though one of them is very similar to the analysis we imposed on the sieve in E (fig. 6). Take any two elements along any diagonal. Can you find a number such that any two elements on any diagonal must differ by a multiple of that number (excluding 1 from the analysis)? Perhaps another glance at figure 4 for a spiral in N will inspire an answer to that question in N. In N, the difference between any two adjacent diagonal elements can never be less than 2, and any two elements differ by some multiple of 2. Sometimes that multiple of 2 is quite large; sometimes it is small. Can you find a number that acts like 2 for E? It appears that 4 fills the bill (see ex. 3.3.16). Given

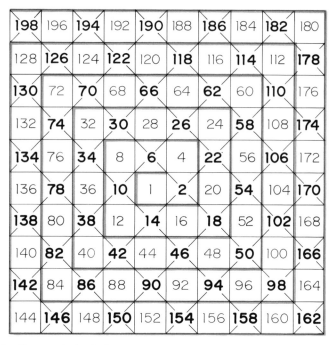

Fig. 7. A spiral grid indicating the distribution of primes in the set E

any element e along a diagonal, any other one is generated by the formula $e + 4k$, for $k \in N$. Once more (as with the sieve) we see why the primal character of any element e along a diagonal tells us something about all the elements on the diagonal.

It's as if you were able to demonstrate that every member of a family is a thief (or a saint) just by knowing about any one member. What a fascinating prospect!

PROBLEM SET 3.3

1. How do we know that any number of the form $2 \cdot \sigma$ for σ any odd element in N must be a prime in E? (We left the matter open in section 2.2.) Well, if it is not a prime, there is some element of E other than 1 and $2 \cdot \sigma$ that divides it. Since we are eliminating 1, we are assured that all possible elements from E must be of the form $2 \cdot n$ for $n \in N$. Suppose there is an element $2 \cdot n$ from E that divides $2 \cdot \sigma$. That is, $2n | 2\sigma$. According to the definition of *divides*, there then must be some element of E (other than 1) of the form $2m$ such that $(2n) \cdot (2m) = 2\sigma$. But then $2nm = \sigma$, and we have an odd number expressed as twice some number, which is impossible. Criticize the proof above. Find a friend and see either if you can find loopholes in it or if you can come up with a better (perhaps simpler)

explanation. You might find it fun to list all the assumptions being made in the proof. Which are questionable?

2. How do we know that if a number is not of the form $2 \cdot \sigma$, then it is not prime? Assuming we are dealing with all elements of E other than 1 (why?), we know that they must be either of the form $2 \cdot even$ or $2 \cdot odd$ (why?). We showed in 3.3 what happens in the second case. Now look at the first case: $2 \cdot even$. It can be expressed as $2 \cdot 2n$ for some $n \in N$. What must divide that number other than 1 and itself?

3. What test can we use to determine primality in T?

4. Create some system of your own that resembles E and T, and consider whether or not you can come up with a test of primality.

5. In section 2.2 we suggested the following two modified Fermat and Mersenne generating formulas for E:

$$\text{Fermat:} \quad 2^{2^n} + 2$$
$$\text{Mersenne:} \quad 2^p - 2$$

Using a test for primality that we developed in section 3.3, show whether or not they generate an infinite number of primes. Which test or tests do you find convenient to use? Determine what difference it makes whether the domain of n and p is N or E.

6. Now that you have a surefire means of testing for primality in E, return to exercise 2.2.7, on modified Euler conjectures in E. The exercise has many parts, but we are essentially concerned with whether or not three functions—$x^2 + x$, x^2, $x^2 - 2$—generate primes in E. In answering the question, we must distinguish what is the domain of x.

7. In section 2.2, we examined $2 + 8n$ and found that despite the fact that 2 and 8 are not relatively prime, $2 + 8n$ generates an infinite number of primes. Prove that this is so using at least two different divisibility tests.

8. Have you come across other mathematical illustrations (not in number theory) of the show tune "With Me It's All or Nothing?" Try not only to answer the question but to see if you can clarify it. What might be several different interpretations of it?

9. We claimed that since there would be greater regularity in E than in N, the results in E would be less interesting. Do you agree with that claim? Isn't mathematics supposed to be about finding patterns and regularity rather than chaos? So how can that assertion be true? This is another nice question for you to debate. We shall raise this question for you again in chapter 6.

10. Can you explain why it is that in E, the first primes sieve out all composites? There is no more work left for other primes. Is this a function

of the form of the sieve? What would happen if you chose a sieve seven boxes wide instead of six? Try it.

11. If you make a sieve nine boxes wide for finding primes in E, how do entries in the same column compare?

12. We have claimed that in using a sieve six boxes wide in E, almost every entry is of the form $a + 12n$ (which entry isn't?), and furthermore, that if a is prime, then so is $a + 12n$; otherwise it is not. This resembles Dirichlet's argument, which we offered in E, except for the fact that $n \in N$ rather than E. Show why the fact that $n \in N$ makes no difference here in applying Dirichlet's argument.

13. Make up a sieve for T (up through 213), again using six columns, and indicate those diagonals along which only primes in T fall.

14. Make a Ulam spiral for elements of T and see what we can say about prime-rich and prime-poor diagonals.

15. On some occasions, two adjacent diagonal elements in the Ulam spiral in E differ by exactly 4; at other times they differ by a medium-sized multiple of 4; at still other times they differ by a very large multiple of 4. When do each of these three conditions occur? Can you find some way to show that any two elements along a diagonal must differ by some multiple of 4?

16. What is the magic number in T that acts like 4 in Ulam's spiral in E?

17. Here is one way of finding $\pi(t)$ for t belonging to T. First let us look at E again. Recall that each element in E (excluding 1) is expressible in one of exactly two ways. Primes are of the form $4n - 2$ ($n \in N$) and composites are of the form $4n$ ($n \in N$). Let us compare this situation with T. In T, each element (excluding 1) is expressible in one of exactly three ways. Each prime is either of the form $9n - 3$ or $9n - 6$ ($n \in N$), and each composite is of the form $9n$.

About two-thirds of the elements of T are prime (why is this true?) and only about one-third of the elements of N appear in T (why is this true?). Therefore a reasonable conjecture at this point is $\pi(t) = \frac{1}{3} \cdot \frac{2}{3} t = \frac{2t}{9}$ for t composite. (Compare this discussion to the development of $\pi(e)$ for e composite in E.) Convince yourself that $\pi(t)$ does in fact equal $\frac{2t}{9}$ for t composite in T. For $t \in T$, t a prime of the form $9n - 3$, $\pi(t) = \frac{2t + 3}{9}$. Verify that these formulas "check out." (Compare them with $\pi(e)$ for e prime in E.) Find a formula for $\pi(t)$ if t is of the form $9n - 6$. Can you now find *one* formula—using the "[ ]" function as we did in *E*—that covers all the different situations in *T*? That is, when *t* is 1; when *t* is prime; when *t* is composite (in two different ways). Find an expression for $\pi(t)$ that works regardless of which of the four conditions is being considered. (This was modified from a problem submitted by George Goodwin, a student from the State University of New York at Buffalo who attended my course in 1978.)

18. Make a "triangular" spiral for the elements of *T* in the following way:

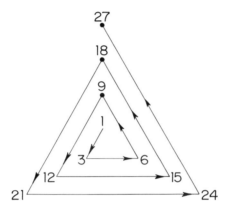

a) Extend the triangular spiral for a few more layers.

b) What can you say about primes and composites in *T* on the basis of this triangular spiral?

(This problem was suggested by Dipendra Bhattacharya, a student from the State University of New York at Buffalo in my course during 1978.)

19. We have found a very simple divisibility test for the primality of elements of *E*. Take any element *e* (not equal to 1) of *E*. If 4 divides *e* in *N*, then *e* is composite. If 4 does not divide *e* in *N*, then *e* is prime.

Though it is easy enough, in *N*, to perform division by 4, there is an interesting shortcut. Take any two-digit number—like 24—that is divisible by 4 in *N*. Now consider any other number that ends in 24. Is that new number still divisible by 4? Take any two-digit number *not* divisible by 4 in *N*, like 18. Now consider any other number that ends in 18. Is that new number divisible by 4? You can now conjecture a simple test for divisibility by 4 for large numbers (those with more than two digits) by looking only at the last two digits. What is it? (This problem is a modified version of one suggested by Philip Fanone, a student from the State University of New York at Buffalo in my 1978 course.)

4

Unique Factorization and Surroundings

A Recipe to Reduce

WE BEGAN our introduction to this book (section 1.1) with the observation that we expect a unique answer when we reduce fractions to lowest terms. So, for example, when we reduced 12/36, we got 1/3, since 12 divides both the numerator and the denominator.

We could, of course, have been a little obtuse and still arrived at 1/3. For example, suppose we had seen at first only that 2 divided both numerator and denominator. We might then have had

$$\frac{12}{36} = \frac{\cancel{2} \cdot 6}{\cancel{2} \cdot 18} = \frac{6}{18}.$$

On the other hand, if we saw that 6 divided both 12 and 36, we might have gotten

$$\frac{12}{36} = \frac{2 \cdot \cancel{6}}{6 \cdot \cancel{6}} = \frac{2}{6}.$$

In neither instance have we finished, for 6 and 18 have some factor greater than 1 in common (thus are not relatively prime), and therefore 6/18 can be further reduced. Since 6 divides both numerator and denominator, we can reduce to 1/3. Similarly, 2 and 6 are not relatively prime, and if we divide them both by their greatest common factor, 2, we get 1/3.

All three paths, then, lead us to the same lowest-term reduction of 1/3. To say that the number has been reduced to lowest terms means that we have found an equivalent expression for 12/36, one in which the numerator and denominator have no common factors other than 1. That is, the numerator and denominator are relatively prime.

At the beginning of the book, we asked you to consider whether things might be different. All our experience suggests not, but do we appreciate the significance of that experience? We shall try to understand the significance of that experience by examining the analogous issue in E. Is it true that when we choose our fractions so that numerators and denominators are from E, we get the same results? Let's try a few cases. Remember that all our elements and operations are now from E.

Let us consider the following: $12/20 = ?$ In E there is only one number that divides 12 and 20: it is 2. (Why does 4 not work?) Therefore

$$\frac{12}{20} = \frac{6 \cdot \cancel{2}}{10 \cdot \cancel{2}} = \frac{6}{10},$$

and reduced to lowest terms, $12/20 = 6/10$. Though $6/10$ may look like a peculiar reduction to lowest terms, it is legitimate because 2 does not divide 6 or 10 in E.

So far, so dull. Let's try another example, $10/24$. The number 10 is prime in E, and since 24 is not a multiple of 10, the fraction is already reduced to lowest terms. Take still another, $16/24$. There are two factors that 16 and 24 have in common, 4 and 2. If we divide numerator and denominator by 2, we get

$$\frac{16}{24} = \frac{\cancel{2} \cdot 8}{\cancel{2} \cdot 12} = \frac{8}{12}.$$

If we divide numerator and denominator by 4, we get

$$\frac{16}{24} = \frac{4 \cdot 4}{6 \cdot 4} = \frac{4}{6}.$$

Since 4 and 6 are relatively prime, we're done. Let's return to $8/12$, however. Can we reduce it further? Yes, for 2 is a factor of both 8 and 12. We therefore get

$$\frac{8}{12} = \frac{\cancel{2} \cdot 4}{\cancel{2} \cdot 6} = \frac{4}{6},$$

and we have the same reduction of $16/24$ to lowest terms that we had in N.

Well, what's new? We have shown that E acts like N with regard to reducing fractions to lowest terms. Look once again at the first problem we selected in this chapter for N. We cannot reduce $12/36$ to $1/3$ in E, for 12 does not divide 36 in E. Let us therefore look at our two "obtuse" solutions. By dividing by 2 we end up with $6/18$, and by dividing by 6 we end up with $2/6$. But look carefully at both solutions. Both are reduced to lowest terms, for their numerators and denominators are relatively prime. Yet they have different numerators and denominators—a phenomenon we did not come across in N.

Thus, some fractions in E can be reduced to lowest terms in such a way that the results are different! After reading the next subsection, you ought to be able to figure out which fractions these are, but perhaps you would like to ponder the question before reading on.

Behind It All

What is the significance of our findings regarding the reduction of fractions? A couple of issues are involved. We shall explore each of them separately in the next two subsections ("Unique Factorization" and "The Fall of the Greatest Common Divisor"), and then, in the following section ("Well, Order!"), we shall look deeper at a root of both. It may be worth returning to this issue after reading chapter 6.

Unique Factorization

Why do we never run into trouble when reducing fractions to lowest terms in N? Look back at the procedure we generally follow. Though we have been a little sloppy in our approach, we could follow a more systematic procedure. Given a fraction in N (one whose numerator and denominator are from N), we could factor both numerator and denominator into products of primes and then just eliminate all duplications from both places. For example, suppose we had to reduce to lowest terms 90/105. We might proceed as follows:

$$90 = 9 \times 10 = 3 \times 3 \times 5 \times 2$$
$$105 = 5 \times 21 = 5 \times 7 \times 3$$

Then,

$$\frac{90}{105} = \frac{3 \times 3 \times 5 \times 2}{5 \times 7 \times 3} = \frac{3 \times \cancel{3} \times \cancel{5} \times 2}{\cancel{5} \times 7 \times \cancel{3}} = \frac{6}{7},$$

and we have reduced to lowest terms.

We could have proceeded to get the prime factorizations differently if we had been so inclined. For example, we might first have observed that

$$90 = 15 \times 6;$$
$$105 = 21 \times 5.$$

Then, further breaking these up into products of primes, we would have gotten

$$90 = 15 \times 6 = 5 \times 3 \times 2 \times 3;$$
$$105 = 21 \times 5 = 7 \times 3 \times 5.$$

Before long, we would have arrived at the same result as above.

Our experience in N tells us that, except for 1, numbers are either primes or composites. If they are composites, then they can be expressed as products of primes. But that is only part of the story, because our experience in E also tells us that numbers are either primes or can be expressed as products of primes. For example, $20 = 10 \cdot 2$; 14 is prime already; $24 = 2 \cdot 2 \cdot 6$.

The other part of the story unfolds when we look at numbers like 36 in E. What we implied in our earlier calculations is that 36 also can be expressed as a product of primes, but something very peculiar happens. The number 36 is $18 \cdot 2$ when expressed as a product of primes, but it also equals $6 \cdot 6$ when expressed as a product of primes. *Thus, in E, some numbers can be factored into a product of primes in more than one way!* Try 60 and 72 in E and see what happens. Can you predict those numbers that will not be uniquely factorable into a product of primes in E? This is an interesting issue, but we shall hold off discussion until the last subsection of this chapter.

Let us pinpoint sharply the property that numbers have in N with regard to products: *All numbers except 1 are either primes or products of primes. Furthermore, if a number is a product of primes, then (with the exception of the order of the factors) there is only one way in which it can be expressed as a product of primes.* This assertion is known as the fundamental theorem of arithmetic, or the unique factorization theorem. We shall say some more about the foundations of this theorem after discussing another breakdown in E.

Before turning to that other breakdown, let us consider one incidental but interesting ramification of the fundamental theorem of arithmetic. We never made very clear why we define primes in N so as to exclude 1. After all, we could define a number as prime if it has at *most* two different factors, and then 1 would be eligible. One reason for excluding 1 as a prime is that we then would have to revise the statement of the fundamental theorem of arithmetic. Why? Suppose you were asked to factor 15 into a product of primes. One person might say 5×3; another could claim it is $1 \times 5 \times 3$; still another might want to factor it as $1 \times 1 \times 5 \times 3$, and so forth. In short, we would approach a situation of the kind we have in E: There would no longer be unique factorization of a composite into a product of primes.

We could, of course, allow 1 to be a prime and then modify the statement of the fundamental theorem of arithmetic; but many other theorems that we have not discussed would then also have to be modified to exclude 1 as a possible prime, which would be a cumbersome task. We turn now to another fundamental property in N that fails in E, bearing on the reduction of fractions.

The Fall of the Greatest Common Factor

If we look carefully at the reduction of 12/36 in E, we notice something else that is significant in our comparison of E and N. For one reduction, we relied on the fact that $2|12$ and $2|36$, displayed by

$$\frac{6 \cdot \cancel{2}}{18 \cdot \cancel{2}} = \frac{6}{18}.$$

For the other reduction, we relied on the fact that $6|12$ and $6|36$, displayed by

$$\frac{\cancel{6} \cdot 2}{\cancel{6} \cdot 6} = \frac{2}{6}.$$

Thus, we saw that there were two different factors of 12 and 36 in E—2 and 6. Of all the factors of 12 and 36 in E, which one is the greatest? (More accurately, the great*er*?) There are no other factors of both 12 and 36 in E, and obviously 6 is the greatest.

In order to hypothesize what is missing in E and present in N, let us look at all the common factors of 12 and 36 in N: 2, 3, 4, 6, and 12 are common factors of 12 and 36. What do you notice about the relationship of all the common factors to their greatest common factor, 12? There are many things to notice, but if you compare the relationship of all common factors with the greatest one in N, and the relationship of all common factors with their greatest in E, a concept relevant to a theme of this exploration might emerge. Notice that in N, all the common factors divide the greatest one! This is not generally true in E. In E, 2 and 6 are factors of 12 and 36; 6 is the greatest one; yet $2 \nmid 6$. We leave as an exercise the exploration of further implications of this "breakdown" (see ex. 4.6). Let us briefly summarize the greatest common factor property in N: If d is the greatest common factor of a and b, and if c is any other common factor, then $c|d$.

Well, Order!

We have already proved (in ex. 2.1.4) half the fundamental theorem of arithmetic in N and have asserted without proof the greatest common factor theorem. Without greater attention to the proofs of both these theorems we cannot carefully examine what accounts for the breakdown of properties in E that hold in N. We shall begin the discussion of structural differences between N and E that might account for breakdowns in analogous proofs in chapter 6. For the moment we shall reexamine an important root of both theorems, for it will explain how we can get at some properties common to N and E that are so obvious that we rarely even notice them. Our focus here will be on N, though the reader may find it enlightening to explore analogies in E.

We know the set N has a least element, but what can we say about any subset of N? Consider, for example, the following:

{1, 3, 7}   {2, 4, 6, 8, ...}
{2, 3, 4, 5, 6, ...}   {x:x > 5}

Each of these subsets has a least element also. What is it in each case? The property that every subset of N has a least element is so obvious that we might not even think it is worthy of notice. One way of demonstrating its significance is to show that not all domains have this property. For example, take the set of positive and negative integers, and give an example to show that not every subset has a least element.

Another set that does not enjoy this property is the set Q of all non-negative fractions. Consider the following subset of Q:

$$S = \left\{x : x > \frac{3}{4}\right\}$$

Is there a least element in S? We might be tempted to answer "3/4," but 3/4 does not belong to S. On the other hand, each of the least elements from the subsets of N (above) does belong to the subset itself.

The property that every subset of N has a least element is called the *well-ordering property*. We shall now make use of that property to prove something we accepted (but questioned) in section 3.1. In trying to find divisors of 3599, we asked how we knew that there were none bigger than 3599—a question that would be hard to answer without machinery of the most fundamental type.

Proof that if $n|a$ in N and $n \neq a$, then $n < a$. Let us begin a proof for the general case. Suppose $n \in N$, and $a \in N$, and we want to show that if $n|a$, then $n < a$ (assuming $n \neq a$). We know a reasonable way to begin. If $n|a$, then by definition there is a $b \in N$ such that $n \cdot b = a$.

Now how are we going to show that $n < a$? In the equation above, the variable n is "hinged" to b by multiplication. If we were to unhinge it to get an equation of the form $k + n = a$ for some $k \in N$, then we would have $n < a$ (though we haven't offered a formal definition, that is how we might define $n < a$).

Can we unhinge n? We want to take $n \cdot b$ and write it as $k + n$. That is, we want $n \cdot b = k + n$. What should k be? Solving for k, we get

$$k = nb - n = n \cdot (b - 1).$$

Therefore, rewrite the original condition, asserting that $n|a$ as follows: If $n|a$, then there is a $b \in N$ such that

$$\underbrace{n \cdot (b - 1)}_{k''} + n = a.$$

Now look at $n \cdot (b - 1)$. We know that $n \in N$, but what about $(b - 1)$? We assumed that $n \neq a$; it then follows that $b > 1$. If $b > 1$, and 1 is the smallest member of N, then $(b - 1)$ belongs to N. Further, $n \in N$, $(b - 1)$

91

ϵN, and therefore, by closure, $n \cdot (b - 1) \epsilon N$. By the definition of the symbol $<$, we can see that $n < a$— and our proof is complete.

Wait! What happened? We said we were going to illustrate the use of the well-ordering principle, yet we did not make use of it at all! Perhaps you would like to question a number of the assumptions we made in the proof above. One assumption, however, is so fundamental (on the order of magnitude of the very divisibility property we are trying to prove) that you may have overlooked it. We assumed that since 1 is the smallest member of N, and $b \epsilon N$ but $b > 1$, that $b - 1 \epsilon N$.

A basic assumption in the proof above is, of course, that 1 is the smallest member of N. How do we know that? How do we know that once we start counting elements of N, we will not reach a point that leads us back to some number smaller than 1? You properly should ask yourself at this point what, specifically, we are supposed to know about N. If we can't answer that, then how can we have come so far? Since this is a reasonable question, let us articulate, in the context of the following proof, at least some of the assumptions we are making, leaving a more thorough examination of our assumptions as an exercise.

We shall now show that 1 is the smallest element of N through a proof by contradiction.

Proof that 1 is the smallest element of N. Assume that 1 is *not* the smallest member of N. There then is some x such that $x \epsilon N$ and $x < 1$. Since there is at least one such x, consider the set of all numbers, the smallest first (if x is the only member of the set, then of course it is the smallest). Call y the smallest member of the set of all xs such that $x < 1$. Then $y < 1$, and y is the smallest such element. But if $y < 1$, we can multiply the inequality by any element of N, and by closure, get another element of N. If we choose the element by which to multiply wisely, we shall have finished the proof. Select the element to be y itself. Then $y \cdot y < 1 \cdot y$, or $y^2 < y$. This obviously contradicts the assumption that y is the smallest element of that subset, and since every subset of N must have a least element, that particular subset (all xs such that $x \epsilon N$ and $x < 1$) must not exist at all.

The proof above most likely appears very elusive. Especially if you have not done that kind of proof before, you ought to have lots of questions. If you are frustrated by this discussion, you might want to bypass it; if you are confused but tantalized, you might want to discuss it carefully with a friend.

Our point has been to demonstrate what kind of machinery is needed to capture the tamest of lambs. The machinery is not only subtle and fascinating, but open to challenge as well. As we implied earlier, some schools of mathematical philosophy would not accept this proof, on several grounds. You might wish to look back at some comments we made about the intuitionist school (Brouwer, Kronecker, and Poincaré) in section 2.1, to see if you can intuit what their criticism might be. You might enjoy doing some reading in the philosophy of mathematics, if for no other reason than to discover that one man's airtight proof is another man's nonsense—not because the "nonsense" man cannot follow the proof, but because he challenges the most basic assumptions that the "airtight" man makes.

After this rather heady discussion and analysis, let us turn, in the last subsection of this chapter, to a further look at the breakdown of unique factorization in E. Though the well-ordering property is an essential ingredient in the proof of unique factorization in N, the fact that unique factorization fails in E is a signal that the well-ordering property is not the only essential ingredient. Why?

How Nonunique Can You Be?

We have shown that in E, 36 can be factored into a product of primes in two different ways. How can we characterize other elements in E for which that is possible? One way of approaching the problem would be to focus on 36. If it is nonuniquely factorable, then you might come up with some reasonable hunches regarding other numbers that derive from 36. You may begin that way if you wish, but if you do, perhaps you should put this book away for a while.

We shall take another tack here. Suppose we multiply two primes in E. We then have a number of the form $(2 \cdot \sigma_1) \cdot (2 \cdot \sigma_2)$ for σ_1 and σ_2 odd in N. Now how can we manipulate that expression to get *another* prime factorization out of it? Keeping in mind that a number in E is prime if it can be expressed in the form $2 \cdot$ odd for odd in N, we need do little more than commute some terms and reassociate the parentheses:

$$(2\sigma_1) \cdot (2\sigma_2) = (2) \cdot (2\sigma_1 \sigma_2)$$

The two terms in parentheses on the right are also primes but different from the terms on the left.

How can we get some additional such examples? Let's see: $(2 \cdot 3) \cdot (2 \cdot 3)$ is 36, and we have already examined how it breaks down. Suppose we take 5 for σ_1 and 3 for σ_2. Then $(2 \cdot 5) \cdot (2 \cdot 3) = 60$. How else can you factor 60? It should now be easy to produce an infinite number of "failures," that is, numbers that flout the prime-factorization theorem.

Can you work backwards, though? Given a number, how could you determine whether it is a failure? It is one thing to come up with a failure on your own and another to recognize one when it is handed to you!

PROBLEM SET 4

1. Reduce the following to lowest terms in E:

    $$\frac{10}{84}; \quad \frac{12}{84}; \quad \frac{20}{36}; \quad \frac{20}{12}.$$

2. Reconsider the sieve of Eratosthenes. We said that sometimes when attempting to cross out a multiple of a prime we find out that it has already been sieved out by an earlier prime. Investigate those circumstances under which you would expect that to happen.

3. Suppose someone claimed to have shown on his calculator that $3 \cdot 7 \cdot 11 \cdot 23$ gives the same answer as $5 \cdot 7 \cdot 13 \cdot 19$. Short of calculating each product, show several different ways in which you might try to verify or refute his claim. Which of the procedures was the easiest?

4. In justifying the definition of a prime in N that excludes 1 as a prime, we have essentially relied upon an aesthetic criterion and not a logical one. What is the aesthetic criterion?

5. People frequently say that definitions are arbitrary in that you can define things any way you want. What do you think they mean by this? Do you think it is true? Could you define a circle as a straight line that goes on forever? How does this question bear on anything we have done in this unit?

6. We found that both 2 and 6 have the property that they divide 12 and 36 in E. Furthermore, 6 is the largest common factor of 12 and 36. Contrary to its behavior in E, 6 is *not* divisible by this other factor.

   a) Can you find other instances in E in which this breakdown occurs?

   b) What is the significance of this breakdown as regards the reduction of 12/36 to lowest terms? Suppose 2 did divide 6 in E; what effect would this have on the reduction of 12/36 to lowest terms?

7. We actually have proved half the fundamental theorem of arithmetic in this book, though this was not our explicit intention at the time. See exercises 2.1.2–4 if you are interested. Show that we proved only half the theorem.

8. In proving that if $n|a$ and $n \neq a$, then $n < a$, we had to show that $b - 1$ belongs to N. Why was it not obvious that if $b \in N$, then $b - 1$ belongs to N? What value of b would have knocked $b - 1$ out of the domain of N?

9. Look again at the proof that 1 is the smallest member of N. What additional hidden assumptions were made in the proof? How impressed were you by the proof? If you were thoroughly confused, what would it take to enlighten you? Are there some aspects of the proof that you understand but do not consider legitimate?

10. Which of the following are nonuniquely factorable in E: 72, 100, a number of the form $4p$ for p a prime in N? Can you come up with a general test of failures in E?

11. How do the failures (numbers not uniquely factorable as a product of primes) in E cluster?

12. How would you prove that there is no member of N between 1 and 2?

13. List three elements of T for which the fundamental theorem of arithmetic holds. List three for which it fails.

14. All our failures in E had two prime factorizations. The following example has three:
$$216 = 6 \cdot 6 \cdot 6 = 2 \cdot 6 \cdot 18 = 2 \cdot 2 \cdot 54$$
a) Find two other numbers that have three prime factorizations in E.
b) How could you characterize all those elements of E that have three different prime factorizations?
c) Try to generalize to four different prime factorizations. (You may check the accuracy of your conjecture, of course, by trying examples to see whether they tend to substantiate or contradict your conjecture.)

15. Can you find a test for a "three-time loser" in T? If not, show why such numbers cannot exist.

16. Look once again at the subsection "Well, Order!" Does E have the well-ordering property? Which theorems in that section are also theorems in E?

5

Odds'n Evens

WE BEGAN our journey with a jolt (in section 1.3) when we pointed out that, provided we are clear about context, there is more than one prime in E. This chapter will provide us with additional surprises (leading to a few more jolts) as we once again clarify some ambiguities about the definitions of *even* and *odd*.

What is more obvious than the concept of odd and even? Well, what are the even numbers? Some of you might say that they are the numbers in E, and others might claim that they are all numbers divisible by 2 (see ex. 5.1 for one possible misconception). If we take as our root concept that even numbers are those that are divisible by 2, then what are the evens in

$$E = \{1, 2, 4, 6, 8, 10, 12, \ldots\}?$$

As with the concept of prime numbers, so with that of evens. In E, 6 is not even, because 2 does not divide 6 in E. The evens, then, are all the composites in E, 4, 8, 12, 16, 20, ..., for 2 divides each of them, although it does not divide any of the primes in E. The numbers that aren't even (and are therefore odd), are either the primes or 1.

There is one mildly interesting application of this broader concept of evenness. Recall that in section 2.2 we asked how many twin primes there were in E and that we had two different interpretations of twin primes—one yielding "none" as an answer and the other "an infinite number." Our more general definition of *even* gives added credence to the answer "infinite number." The reason is that with the exception of 2 in N, there are no even primes, so twin primes are investigated with regard to consecutive odd elements of N. If we search for twin primes in E from the same perspective, then the next odd element after any prime $2e - 2$ would have to be $2e + 2$.

Let us now explore another avenue—called parity—that relates evenness and oddness in a way that is more than mildly interesting. In 1742, Goldbach came up with a fascinating, easily understood conjecture that has eluded proof or disproof for over two hundred years. As you look at the conjecture, you might do some pseudohistory of your own and ask how, most likely, he came up with it (see ex. 5.3).

He noticed that if he started with an even number, he could "decompose" it into a sum of two primes. So, for example,

$$4 = 2 + 2$$
$$6 = 3 + 3$$
$$8 = 5 + 3$$
$$10 = 5 + 5$$
$$20 = 3 + 17.$$

He was hoping to prove that this could be done for any even number (with one obvious exception), but he was able neither to prove it nor to disprove it by finding a counterexample. Furthermore, though no proof or disproof exists today, there has been headway made on the problem. You might wonder how one can make headway if one cannot furnish proof. One strategy is to collect a great deal of evidence in favor of the idea, as Ulam did. Another strategy would be to offer a limited proof—one, for example, that covers only part of the territory, or one that makes more assumptions than we desire.

The headway was made in 1931 by a Russian mathematician, Schnirelman. His generalization is quite humorous, considering what Goldbach had hoped to prove. Goldbach believed that you need no more than two primes (though you cannot get away with fewer either) to express any even number as the sum of primes.

If you were unable to prove that you could write any even number as a sum of two primes, you might be willing to lower your sights a little bit and try to prove that you need three or four primes for any even number. Perhaps you'd stretch your imagination and try to show that you need no more than half a dozen primes.

But Schnirelman came along and proved that you do not need more than 300 000 primes to represent any even number as a sum of primes! Though this was certainly progress, can you image the belly laugh Goldbach would have had if he had heard of such a proof?

There is an interesting sidelight on Goldbach's conjecture that is worth examining before we do the inevitable—search for analogies in E. Some even numbers can be represented as a sum of primes in exactly one way. For example, 8 can be written only as $3 + 5$. Some numbers are more "open," however:

$$18 = 5 + 13, \text{ but } 18 \text{ also equals } 7 + 11;$$
$$20 = 17 + 3, \text{ but it also equals } 13 + 7.$$

With a little imagination, we might call numbers that can be represented as a sum of two primes in two different ways "Copperbach numbers." For suggestions for further exploration of this sidelight, see ex. 5.6 and 5.7.

Let us now explore what happens in E. Can any even number be represented as the sum of two primes? Let's try a few:

$$6 = ?$$
$$8 = ?$$
$$10 = ?$$
$$12 = ?$$

Can you find four sets of two primes in E whose sums produce the even numbers above?

Since we know that the pairs on the right-hand side must come from E, let us list all the possible ways of adding pairs from E to arrive at the indicated number:

(i) $6 = 2 + 4$
(ii) $8 = 2 + 6$
(iii) $10 = 4 + 6 = 2 + 8$
(iv) $12 = 4 + 8 = 6 + 6 = 2 + 10$

In (ii), we have truly expressed 8 as the sum of two primes. In (iv), two out of the three examples work (which does not?). Examples (i) and (iii), however, are no good, for neither 4 nor 8 are prime. It looks as if we have a breakdown in Goldbach's conjecture before we even get started.

Does Goldbach's conjecture break down in E? A careful glance at (i)–(iv) above will remind us that we have overstepped our bounds, so to speak. How so? We have not selected the even elements in E in all cases for (i)–(iv)! Specifically, 6 and 10 are not even in E, as we discussed in the beginning of this chapter. It looks, so far, as if things might work out satisfactorily if we select the evens in E.

Perhaps we can get a clue from looking at 8. We can easily see that $8 = 2 + 6$, where both 2 and 6 are prime. Though there were two ways of representing 12 that worked, one of those two ways fits the same pattern as 8. That is, $12 = 2 + 10$. Can we find a continuation of that pattern for the next even? What does 16 equal as the sum of two primes? Continuing the pattern, $16 = 2 + 14$.

It looks like we're in! Can we always represent an even number in E as the sum of two primes in E? If we select 2 as one of the primes, automatically the other seems to be prime as well. Will that always be so? To answer that question we need some minor machinery. There are several ways of seeing what is going on, but the simplest might be as follows: Suppose we start with any even number of E. It must be of the form $4j$ for $j \in N$. Then, if we represent $4j$ as the sum of 2 and some other number, will that other number necessarily be prime in E? Let's see: if we call the other number x, then $4j = 2 + x$, so that $x = 4j - 2$. But obviously $4j - 2$ is prime in E—and we're done!

Look what has happened here. A problem that has eluded the grasp of research mathematicians for over two hundred years curls up in embarrassment before most junior high school youngsters when we shift the context from N to E. Problems that are tigers in one context sometimes become lambs in another, slightly different context. Where have you come across that theme before?

The exercises will leave you with a number of challenging explorations of Goldbach's conjecture in different contexts, but before turning to them, you might enjoy seeing whether or not the failure of (i) and (iii) was a fluke. That is, why can't a prime in E be expressed as the sum of two primes? Perhaps it would be possible to do so for some primes, though not for 6 and 10.

To see what is going on, suppose you add any two primes in E. What must you get? Let's try it: If one prime is $4j - 2$ and another $4k - 2$, then

$$(4j - 2) + (4k - 2) = 4(k + j) - 4 = 4(k + j - 1).$$

But what kind of number is $4 \cdot (k + j - 1)$? Being of the form $4 \cdot m$, it is even in E and therefore *not* a prime. So a prime in E (like 6, 10, 18, 102) can never be expressed as the sum of two primes. But that is not distressing, since all the primes in E are odd in that domain, and Goldbach's conjecture applies only to even numbers in N. The analogy therefore holds.

Though at first the concept of parity seems either trivial or inappropriate in E, on reflection we can see that it applies with the same force as it does in N, and in doing so it enables us to find a meaningful analogy for a conjecture that has eluded mathematicians (working in N) for two centuries (see ex. 5.13).

It is worth stressing that the concept of parity is a fundamental and exciting one in mathematics, and we have done little more than scratch the surface in one very limited area. If you are interested in other applications of the notion of parity, you might explore some popularized works in topology. The Königsberg bridge problem and variations of it show how the notion of parity can shed light in places where one would least expect it to have relevance.

PROBLEM SET 5

1. Some people would claim that even numbers are those that end in 0, 2, 4, 6, 8. A problem with this definition is that it links the concept of

evenness to base-ten numbers. How could you tell what numbers were even in base two? In base three? (See ex. 3.1.21 if this question is unclear.)

2. To persuade some people that they are wrong if they want to call 6 even in E, suggest an argument by analogy that would allow you to label an odd number in N even.

3. How do you imagine Goldbach came up with his conjecture that in N, every even number greater than 2 is a sum of two primes? (Though no one knows for sure how Goldbach arrived at his conjecture, you might consider its converse as a starting point for this problem.)

4. In Goldbach's conjecture in N, it turns out that there is one even number that cannot be expressed as the sum of two primes. Show why that counterexample could be used as evidence for defining 1 in N as a prime number.

5. Schnirelman's proof of Goldbach's conjecture covers ground that is too *general* to interest us. What kind of statement might cover an infinite number of cases but still be more *specific* than is useful to us?

6. We found some "Copperbach numbers" in N—numbers that could be represented as the sum of two primes in two different ways. Can you find some more? Can you say anything about how they cluster?

7. Do "Silverbach numbers," that is, numbers that can be written as the sum of two primes in three different ways, exist in N?

8. To stretch our intuition and imagination, let us consider Goldbach's conjecture in the domain T. Our problem here begins in the very statement of the conjecture. Since 2 does not belong to T, there are apparently no even numbers in the set at all, and we thus appear to be in a position to discard the problem completely. Observe, however, that in both N and E the element 2 happens to be the second-smallest element. As for T, the second-smallest number is 3. If we choose this property to define evenness, then in T, any element divisible by 3 is "even." What are the first few even numbers in T? Characterize all even elements of T.

9. The first few primes in T are 3, 6, 12, 15, 21, 24, 30, 33. It is obvious that any element of the form $3t - 3$ for t belonging to T' is prime in T (though such a formula does *not* characterize all primes). With this somewhat stretched interpretation of *even* in T, state and prove Goldbach's conjecture in T.

10. Are there Copperbach and Silverbach numbers in T?

11. The fundamental theorem of arithmetic in N claims that any number can be represented as the product of primes in exactly one way (if we

ignore the question of order). Comment on the state of affairs if we replace *product* by *sum*.

12. Considering Goldbach's conjecture with regard to E we find some interesting things occurring. The *first* pair of even numbers in E, 4 and 8, can be expressed in exactly one way as a sum of two primes in E ($4 = 2+2, 8 = 2+6$).

   a) Consider the *second* pair of even numbers in E, 12 and 16. In how many ways may each of these be expressed as a sum of two primes in E?
   b) Consider the *third* pair of even numbers in E; in how many different ways may each of them be expressed as the sum of two primes?
   c) Can you generalize this situation? (That is, is there a way of determining in how many ways a given even number in E may be expressed as the sum of two primes in E?)
   d) Observe that when 16 is written as all possible sums of pairs of primes, all primes less than 16 are used at least once. Is this generally true? Try some others such as 40.

(This problem is compliments of Wallace Jewell, a student in my course at State University of New York during 1978.)

13. In section 1.2 we explored properties of closure for N, E, and E' under addition and multiplication. Reexamine the subsets of even and odd elements of these three domains for closure under these same two operations. Do the same for T and T'. How much of this exercise did you already investigate in exercises 1.2.3 and 1.2.5?

6

Epilogue

AFTER having come this far, you might find a rereading of the introductory remarks in section 1.1 enlightening. We spoke of the value of changing perspective in order to understand the intellectual significance of an idea. You might now find it profitable to keep those remarks in mind as you take a step back to survey the terrain you have explored.

You have found that some properties or problems in one context (N) are very different in another (E). On most occasions the properties of N were much more unpredictable and lacking in regularity and clear patterns than the properties of E. In addition, both conjecturing and proving were generally harder in N than in E.

We discovered all this in comparing the two domains with regard to—

- determining the number of primes;
- finding a generating formula for primes;
- figuring out if a given number is prime;
- finding a way to predict how primes are distributed in an interval;
- exploring Goldbach's conjecture.

There were some surprises in E, however. In chapter 4 you found out, for example, that the analogue of the fundamental theorem of arithmetic is false in E, and that as a consequence there exists more than one way of reducing some fractions to lowest terms—a real eye opener for anyone who believed that getting a unique answer to such problems is a trivial kind of enterprise.

We hope that this book will inspire you to search further in comparing domains, for many other surprises are in store for you along the way.

Though some of the readings in the Bibliography explicitly make comparisons like those we have made in this book, most standard texts on number theory focus only on N. As you become comfortable with the concepts we have introduced here, we urge you to examine some of those other standard number-theory texts. We suggest you explore on your own how new concepts introduced in those books fare in some of the settings other than N that we have examined in this book.

In Long's book (1965), for example, you will find chapters on multiplicative number-theoretic functions and congruences—topics not covered explicitly in this book. The author develops some elegant theorems dealing with the concept of congruence, but the basic idea is very simple and provides nice analogies with solutions of equations. We shall explore the topic briefly in order to gain some further insight on the explanation of differences between N and E.

We say that a is *congruent* to b modulo c if c divides $a - b$. The notation used to express the definition above would be $a \equiv b \pmod{c}$ provided $c|(a-b)$.

Let us now look at one elementary concept regarding the idea of congruence as it applies to N and E. You know that with equations, if $x = y$ and $y = z$, then $x = z$; this is called the transitive property. In N, we can prove that the same property holds for congruence. For example, $20 \equiv 12 \pmod{4}$ for $4|(20-12)$, and $12 \equiv 8 \pmod{4}$ for $4|(12-8)$. It is therefore also true that $20 \equiv 8 \pmod{4}$ for $4|(20-8)$.

At this point you have all the machinery you need to prove that in general, if $a \equiv b \pmod{c}$ and $b \equiv d \pmod{c}$, then $a \equiv d \pmod{c}$.

Let us look at the analogy in E, however. Let us use the same example we used in N:

$$20 \equiv 12 \pmod{4} \quad \text{Why?}$$
$$12 \equiv 8 \pmod{4} \quad \text{Why?}$$

But 20 is not congruent to 8 (mod 4), because in E, 4 does not divide $20 - 8$. So in E congruence is not always transitive.

Why does transitivity for congruence break down in E? We shall not answer that question directly but shall instead provide you with some fundamental insights on the nature of N versus E that should enable you not only to analyze this question but to understand, as you study standard texts, other points of disparity between the two domains.

Using dots in section 1.2, we introduced an idea that was applied to problems throughout this text:

If $a|b$ and $a|c$, then $a|(b+c)$.

This theorem (as well as variations of it) was applied in exercises in P.S. 3.1 to determine primality in N. Let us now look at the analogue in E. Does it hold? In some cases, yes. For example, if $2|4$ and $2|8$, then 2 does divide

$4 + 8$ as well. In some cases, however, we are in for a surprise. For example, $2|2$ and $2|8$, yet 2 does *not* divide $2 + 8$ in E!

Why does the theorem fail in E? Why do we have some counterexamples? Let us try to prove that $2|2$ and $2|8$ implies $2|(2 + 8)$ and see where there is a breakdown:

$2|2$ because there is a number 1 such that $2 \cdot 1 = 2$.
$2|8$ because there is a number 4 such that $2 \cdot 4 = 8$.

Since $2 \cdot 1 = 2$ and $2 \cdot 4 = 8$, let us add the two equations:

$$2 + 8 = 2 \cdot 1 + 2 \cdot 4$$

If we now employ the distributive principle, we have

$$2 \cdot 1 + 2 \cdot 4 = 2 \cdot (1 + 4) = 2 \cdot 5,$$

and we can claim that $2|(2 + 8)$ because we have a number, 5, such that $2 \cdot 5 = 2 + 8$.

We can see immediately, however, what is wrong with our last equation. Since 5 is not an element of E, 2 does not divide $2 + 8$ in E. If we backtrack a little, we find the root of our difficulty: *The distributive principle is false in E.* That is, consider once more the following: $2 \cdot 1 + 2 \cdot 4 = 2 \cdot (1 + 4)$. Clearly the left side has all elements and operations that are defined in E. On the right, however, we have an element $(1 + 4)$ that does not belong to E at all!

There are instances in which the distributive property applies, but there are also some in which it fails. Going back one more step, we see that when failure occurs, it does so precisely because of a lack of closure under addition in E—an issue we discussed in section 1.2.

Where does all this lead us? The discussion above suggests an important structural difference between N and E. Because of the lack of closure under addition in E, the distributive principle sometimes fails in E. Therefore, as you look at disparities between conjectures in N and E, you might try to identify those instances in which the distributive principle is applied in the proofs of standard texts.

If the distributive principle is essential to a proof in N and if the principle fails in analogous instances in E, then we would expect a true conjecture in N to have a false analogue in E. It would be a worthwhile exercise to examine the proof of uniqueness for the fundamental theorem of arithmetic in N in a standard number-theory text and see the role of the distributive principle.

It is worth stressing that since E is a subset of N, E "inherits" many properties from N. For example, the commutative and associative properties hold for both multiplication and addition in both domains. The number 1 plays the same role under multiplication in both domains. Closure of a subset is, however, a basic property that is not inherited automatically for any subset of a set under a specified operation, and though E is closed under multiplication, its failure to be closed under addition accounts for an important structural difference (the failure of the distributive principle) that affects the validity of some analogous conjectures in N and E.

In closing, we urge you once again to take a panoramic view of N and E as well as of other domains we have explored. Sometimes there were regularity, pattern, and predictability in N, and sometimes we found these attributes in E. You might find it worthwhile to meet with a friend who has gone through this material and to discuss what role these attributes play in your appreciation of mathematical thought. To what extent do you derive pleasure from regularity and expectation, and to what extent are you motivated by surprise, absence of pattern, and unpredictability? These are fundamental questions that all of us need to ask ourselves. They are questions for which there is no right answer, but whatever our answer, the exploration can lead us to a better understanding of ourselves not only as mathematical thinkers—teachers or students—but as people who try to tame an exciting but puzzling world.

BIBLIOGRAPHY

Brown, Stephen I. "Of 'Prime' Concern: What Domain?" *Mathematics Teacher* 58 (May 1965):402-7.

──────. " 'Prime,' 'Elementary,' and 'Fundamental' Comparisons." *Pentagon* 26 (Spring 1967):95-105.

──────. " 'Prime' Pedagogical Schemes." *American Mathematical Monthly* 75 (June-July 1968):660-64.

Gardner, Martin. "Mathematical Games." *Scientific American*, March 1964, pp. 120-28 and cover.

Greenleaf, Newcomb, and Robert J. Wisner. "The Unique Factorization Theorem." *Mathematics Teacher* 52 (December 1959):600-603.

Hawkins, David. "Mathematical Sieves." *Scientific American*, December 1958, pp. 105-12.

Jacobson, Bernard, and Robert J. Wisner. "Matrix Number Theory: An Example of Non-Unique Factorization." *American Mathematical Monthly* 72 (April 1965): 399-402.

Long, Calvin T. *Elementary Introduction to Number Theory*. Boston: D. C. Heath & Co., 1965.

Mills, W. H. "A Prime-Representing Function." *American Mathematical Society Bulletin* 53 (1947):604.

Samuel, Pierre. "Unique Factorization." *American Mathematical Monthly* 75 (November 1968):945-52.

Stein, M. L., S. M. Ulam, and M. B. Wells. "A Visual Display of Some Properties of the Distribution of Primes." *American Mathematical Monthly* 71 (May 1964): 515-20.

Stone, Edward J. "New Domains." *Mathematics Teacher* 58 (October 1965):514-17.

FOR FURTHER READING

Brown, Stephen I. "From the Golden Section and Fibonacci to Pedagogy and Problem Posing." *Mathematics Teacher* 69 (March 1976):180-86.

──────. "Discovery and Teaching a Body of Knowledge." *Curriculum Theory Network* 5 (April 1976):191-218.

Courant, Richard, and Herbert Robbins. *What Is Mathematics?* New York: Oxford University Press, 1959.

Eves, Howard. *Introduction to the History of Mathematics*. New York: Holt, Rinehart & Winston, 1964.

Hamming, R. W. "Impact of Computers." *American Mathematical Monthly* 72 (February 1965):1-7.

Nagel, Ernest, and James Newman. *Gödel's Proof*. New York: New York University Press, 1958.

Perry, William G. *Forms of Intellectual and Ethical Development in the College Years*. New York: Holt, Rinehart & Winston, 1972.

Walter, Marion, and Stephen I. Brown. "Problem Posing and Problem Solving: An Illustration of Their Interaction." *Mathematics Teacher* 70 (January 1977):4-13.